Energy Efficient Hardware-Software Co-Synthesis Using Reconfigurable Hardware

CHAPMAN & HALL/CRC
COMPUTER and INFORMATION SCIENCE SERIES

Series Editor: Sartaj Sahni

PUBLISHED TITLES

ADVERSARIAL REASONING: COMPUTATIONAL APPROACHES
TO READING THE OPPONENT'S MIND
Alexander Kott and William M. McEneaney

DISTRIBUTED SENSOR NETWORKS
S. Sitharama Iyengar and Richard R. Brooks

DISTRIBUTED SYSTEMS: AN ALGORITHMIC APPROACH
Sukumar Ghosh

ENERGY EFFICIENT HARDWARE-SOFTWARE
CO-SYNTHESIS USING RECONFIGURABLE HARDWARE
Jingzhao Ou and Viktor K. Prasanna

FUNDEMENTALS OF NATURAL COMPUTING: BASIC
CONCEPTS, ALGORITHMS, AND APPLICATIONS
Leandro Nunes de Castro

HANDBOOK OF ALGORITHMS FOR WIRELESS
NETWORKING AND MOBILE COMPUTING
Azzedine Boukerche

HANDBOOK OF APPROXIMATION ALGORITHMS
AND METAHEURISTICS
Teofilo F. Gonzalez

HANDBOOK OF BIOINSPIRED ALGORITHMS
AND APPLICATIONS
Stephan Olariu and Albert Y. Zomaya

HANDBOOK OF COMPUTATIONAL MOLECULAR BIOLOGY
Srinivas Aluru

HANDBOOK OF DATA STRUCTURES AND APPLICATIONS
Dinesh P. Mehta and Sartaj Sahni

HANDBOOK OF DYNAMIC SYSTEM MODELING
Paul A. Fishwick

HANDBOOK OF PARALLEL COMPUTING: MODELS,
ALGORITHMS AND APPLICATIONS
Sanguthevar Rajasekaran and John Reif

HANDBOOK OF REAL-TIME AND EMBEDDED SYSTEMS
Insup Lee, Joseph Y-T. Leung, and Sang H. Son

HANDBOOK OF SCHEDULING: ALGORITHMS, MODELS,
AND PERFORMANCE ANALYSIS
Joseph Y.-T. Leung

HIGH PERFORMANCE COMPUTING IN REMOTE SENSING
Antonio J. Plaza and Chein-I Chang

INTRODUCTION TO NETWORK SECURITY
Douglas Jacobson

PERFORMANCE ANALYSIS OF QUEUING AND COMPUTER
NETWORKS
G. R. Dattatreya

THE PRACTICAL HANDBOOK OF INTERNET COMPUTING
Munindar P. Singh

SCALABLE AND SECURE INTERNET SERVICES AND
ARCHITECTURE
Cheng-Zhong Xu

SPECULATIVE EXECUTION IN HIGH PERFORMANCE
COMPUTER ARCHITECTURES
David Kaeli and Pen-Chung Yew

VEHICULAR NETWORKS: FROM THEORY TO PRACTICE
Stephan Olariu and Michele C. Weigle

Energy Efficient Hardware-Software Co-Synthesis Using Reconfigurable Hardware

Jingzhao Ou
XILINIX
San Jose, California, U.S.A.

Viktor K. Prasanna
University of Southern California
Los Angeles, California, U.S.A.

CRC Press
Taylor & Francis Group
Boca Raton   London   New York

CRC Press is an imprint of the
Taylor & Francis Group, an **informa** business

A CHAPMAN & HALL BOOK

Chapman & Hall/CRC
Taylor & Francis Group
6000 Broken Sound Parkway NW, Suite 300
Boca Raton, FL 33487-2742

First issued in paperback 2017

© 2010 by Taylor and Francis Group, LLC
Chapman & Hall/CRC is an imprint of Taylor & Francis Group, an Informa business

No claim to original U.S. Government works

ISBN 13: 978-1-138-11280-3 (pbk)
ISBN 13: 978-1-58488-741-6 (hbk)

Library of Congress Cataloging-in-Publication Data

Ou, Jingzhao.
    Energy efficient hardware-software co-synthesis using reconfigurable hardware / Jingzhao Ou, Viktor K. Prasanna.
        p. cm. -- (Chapman & Hall/CRC computer and information science series)
    Includes bibliographical references and index.
    ISBN 978-1-58488-741-6 (hardcover : alk. paper)
    1. Field programmable gate arrays--Energy consumption. 2. Field programmable gate arrays--Design and construction. 3. Adaptive computing systems--Energy consumption. 4. System design. I. Prasanna Kumar, V. II. Title. III. Series.

TK7895.G36O94 2009
621.39'5--dc22   2009028735

Visit the Taylor & Francis Web site at
http://www.taylorandfrancis.com

and the CRC Press Web site at
http://www.crcpress.com

To Juan, my dear wife.

Thanks a lot for your love!

−− Jingzhao Ou

To my parents.

−− Viktor K. Prasanna

Contents

List of Tables

List of Figures

Acknowledgments

During my Ph.D. study at the University of Southern California, I was working in Professor Prasanna's research group, also known as the *P-Group*. The pleasant research atmosphere in the *P-Group* was very helpful and enjoyable. I would like to thank all the members in the group, including Zachary Baker, Amol Bakshi, Seonil Choi, Gokul Govindu, Bo Hong, Sumit Mohanty, Gerald R. "Jerry" Morris, Jeoong Park, Neungsoo Park, Animesh Pathak, Ronald Scrofano, Reetinder Sidhu, Mitali Singh, Yang Yu, Cong Zhang, and Ling Zhuo. Among them, I want to give special thanks to Amol, who shared an office with me since I joined the group; Seonil, with whom I have discussed many of the research problems; Govindu, who was my officemate for one year; and Bo and Yang, who are very close personal friends in my life.

I would like to thank my colleagues at Xilinx, Inc. This includes Ben Chan, Nabeel Shirazi, Shay P. Seng, Arvind Sundararajan, Brent Milne, Haibing Ma, Sean Kelly, Jim Hwang, and many others. Especially, I would like to thank Brent Milne and Jim Hwang for offering me the internship and the full-time job opportunity at Xilinx. Xilinx is a very innovative company. In the past three years there, I have filed over 30 invention disclosures and patent applications with the company. You can imagine the excitement and fun I have while working there!

I want to express my wholehearted gratitude to my wife, Juan. She has been a great support for me since the first day we fell in love with each other. We were classmates for many years and studied together at the South China University of Technology and the University of Southern California. We are now happily living together in the Bay area with our little angel, Hellen.

Finally, I appreciate the patience, encouragement, and help from Bob Stern and Theresa Delforn at CRC Press.

—— Jingzhao Ou

Preface

The integration of multi-million-gate configurable logic, pre-compiled heterogeneous hardware components, and on-chip processor subsystems offer high computation capability and exceptional design flexibility to modern reconfigurable hardware. Moshe Gavrielov, the recently appointed president and CEO of Xilinx, which is the largest manufacturer of commercial reconfigurable hardware, pointed out that "FPGAs are becoming more and more relevant to a wider range of designers. But it's not just about gates; it's about the IP as well."

Being the leader of the largest programmable logic company, Gavrielov sees a three-fold challenge for his company.

> First, it must maintain its pace in enlarging the capabilities of the underlying silicon. Second, it must build its portfolio of IP across a growing breadth of applications. And third, the company must continue to pour investment into its development tools, so they are able both to serve the needs of an increasingly diverse and, one suspects, increasingly specialized and FPGA-naive community of users and to continue hiding the growing complexity of the actual FPGA circuitry from those users. ... In a semiconductor industry of at least temporarily diminishing expectations, in which much of the growth is sought in the low-power, low-margin consumer world, one that has been firmly resistant to FPGA technology in the past, that is a bold bet. The investment to create new silicon, new tools, and greater scalability will have to come first, and the answer as to whether the growth is really there will come second.*

Rapid energy estimation and energy efficient application synthesis using these hardware devices remains a challenging research topic. Energy dissipation and efficiency have become a hurdle that prevents the further widespread use of FPGA devices in embedded systems, where energy efficiency is a key performance metric. The major challenges for developing energy efficient applications using FPGAs are described in the following paragraphs.

- *The ever increasing design complexity makes low-level design flows and techniques unattractive for the development of many complicated systems.*

---

*From Ron Wilson, "Moshe Gavrielov looks into the future of Xilinx and the FPGA industry," EDN.com, January 7, 2008, http://www.edn.com/blog/1690000169/post/1320019732.html.

With FPGAs being used in many complicated systems, describing these systems using traditional register-transfer and gate-level ("low-level") design flow can turn out to be time consuming and unattractive in many design cases. Synthesis and placing-and-routing of a low-level design typically take up to a couple of hours for a design consisting of multi-millions of logic gates. Even with technologies such as incremental synthesis and module-based low-level implementation, each minor change to the system could incur a significant amount of time to re-generate the low-level implementations. The time-consuming process of generating low-level implementations on FPGA devices prohibit efficient design space exploration and performance optimization.

When we consider the development of digital signal processing and MVI (multi-media, video and image) processing systems, the traditional low-level design flows mentioned above can significantly hamper the communication between hardware designers and algorithm developers. For instance, people from the signal processing community are usually not familiar with hardware description languages, while it is very demanding for a hardware designer to have a profound knowledge of various complicated digital signal processing and image processing algorithms. It is highly desired that domain-specific design tools be developed to bridge the communication gap between the hardware designers and algorithm developers.

- *State-of-the-art design and simulation techniques for general-purpose processors are inefficient for exploring the unprecedented hardware-software design flexibilities offered by reconfigurable hardware.*

From the design description perspective, register-transfer level and gate level techniques are inefficient for constructing reconfigurable platforms containing both hardware and software components. A true system-level design tool is required to describe the bus interfaces and connections between the hardware and software components and to associate software drivers with the hardware peripherals. Xilinx offers Platform Studio [97] for quickly constructing embedded processor systems on reconfigurable systems. Similarly, Altera offers SOPC (System on Programmable Chip) Builder [3] for their Nios processor systems. However, to develop customized hardware peripherals, users still need to reply on the traditional HDL based design flows. As we show later in this book the register transfer level simulation of a processor subsystems is a very time-consuming process and thus unattractive for practical software development.

From the simulation perspective, software programs usually execute for millions of clock cycles, which is far beyond the capabilities of even the fastest industrial RTL (Register Transfer Level) level simulators. The traditional approach for simulating the execution of software programs running on processors is through instruction-set simulators. Many academic and industrial instruction set simulators have been developed for various types of general-purpose processors. Examples of these instruction set simulators include SimpleScaler [14] and Amulator for StrongARM processors [7]. These simulators assume that the targeted processors have a relatively "fixed" hardware architecture. Some configuration options, such as cache size and associativities, memory sizes, etc., are usually available for these simulators. However, as is analyzed in Chapter 5, based on the assumption of relative "fixed" processor architectures, instruction-level simulators are unable to simulate the customized instructions and hardware peripherals attached to the processors. Hence, they are not suitable for hardware-software co-design.

• *Rapid energy estimation and energy performance optimization are challenging for systems using reconfigurable hardware.*

While energy estimation using low-level simulation can be accurate, it is time consuming and can be overwhelming considering the fact that there are usually many possible implementations of an application on FPGAs. Especially, the low-level simulation based energy estimation techniques are impractical for estimating the energy dissipation of the on-chip pre-compiled and soft processors. As we show in Chapter 5, simulating ∼2.78 msec wall-clock execution time of a software program using post place-and-route simulation model of a state-of-the-art soft processor would take around 3 hours. The total time for estimating the energy dissipation of the software program is around 4 hours. Such an energy estimation speed prohibits the application of such low-level technique for software development considering the fact that many software programs are expected to run for tens and thousands of seconds.

On the other hand, the basic elements of FPGAs are look-up tables (LUTs), which are too low-level an entity to be considered for high level modeling and rapid energy estimation. It is not possible for a single high level model to capture the energy dissipation behavior of all possible implementations on FPGA devices. Using several common signal processing operations, Choi et al. [17] show that different algorithms and architectures used for implementing these common operations on reconfigurable hardware would result in significantly different energy dissipations. With this observation, a domain-specific energy performance modeling technique is proposed in [17]. However, to our best knowledge, there is no system-level development framework that integrates this modeling technique for energy efficient designs.

Considering the challenges discussed above, this manuscript makes the following four major contributions toward energy efficient application synthesis using reconfigurable hardware.

• *A framework for high-level hardware-software application development*

Various high-level abstractions for describing hardware and software platforms are mixed and integrated into a single and consistent application development framework. Using the proposed high-level framework, end users can quickly construct the complete systems consisting of both hardware and software components. Most importantly, the proposed framework supports co-simulation and co-debugging of the high-level description of the systems.

By utilizing these co-simulation and co-debugging capabilities, the functionalities of the complete systems can be quickly verified and debugged without involving their corresponding low-level implementations.

- *Energy performance modeling for reconfigurable system-on-chip devices and energy efficient mapping for a class of application*

An energy performance modeling technique is proposed to capture the energy dissipation behavior of both the reconfigurable hardware platform and the target application. Especially, the communication costs and the reconfiguration costs that are pertinent to designs using reconfigurable hardware are accounted for by the energy performance models. Based on the energy models for the hardware platform and the application, a dynamic programming based algorithm is proposed to optimize the energy performance of the application running on the reconfigurable hardware platform.

- *A two-step rapid energy estimation technique and high-level design space exploration*

Based on the high-level hardware-software co-simulation framework proposed in the above section, we present a two-step rapid energy estimation technique. Utilizing the high-level simulation results, we employ an instruction-level energy estimation technique and a domain-specific modeling technique to provide rapid and fairly accurate energy estimation for hardware-software co-designs using reconfigurable hardware. Our technique for rapid energy estimation enables high-level design space exploration. Application developers can explore and optimize the energy performance of their designs early in the design cycle, before actually generating the low-level implementations on reconfigurable hardware.

- *Hardware-software co-design for energy efficient implementations of operating systems*

By utilizing the "configurability" of soft processors, we delegate some management functionalities of the operating systems to a set of hardware peripherals tightly coupled with the soft processor. These hardware peripherals manage the operating states of the various hardware components (including the soft processor itself) in a cooperative manner with the soft processor. Unused hardware components are turned off at the run-time to reduce energy dissipation. Example designs and illustrative examples are provided to show that a significant amount of energy can be saved using the proposed hardware-software co-design technique.

Chapter 1

Introduction

1.1 Overview

Since the introduction of Field-Programmable Gate Arrays (FPGAs) in 1984, the popularity of reconfigurable computing has been growing rapidly over the years. Reconfigurable hardware devices with multi-million gates of configurable logic have become available in recent years. Pre-compiled heterogeneous hardware components are integrated into reconfigurable hardware, which offer extraordinary computation capabilities. To improve design flexibilty and ease-of-use, modern FPGA devices may contain on-chip configurable processor subsystems. These processor subsystems provide great design flexibilities similar to general-purpose processors and programmable DSP (Digital Signal Processing) processors while allowing application optimization. Reconfigurable hardware offers many unique advantages compared with traditional ASICs (Application-Specific Integrated Circuits). The advantages of application development using reconfigurable hardware are discussed in detail in the following sections.

- *Reduced development tool costs*

Designs using FPGAs can greatly reduce the development tool costs compared with application specific ICs [12]. For ASIC based application development, the average seat of development tools costs around $200,000 per engineer. In addition, engineers have to go through a lengthy low-level design process. The low-level design process includes HDL (Hardware Description Language) based register transfer level (RT level) design description, functional and architectural simulation, synthesis, timing analysis, test insertion, place-and-route, design verification, floorplanning, and design-rule checking (DRC). Due to the complicated development process, ASIC engineers often need to use the tools from multiple EDA (Electronic Design Automation) vendors, which dramatically increases the costs for development tools. In comparison, for FPGA based application development, the average seat of development tools costs between $2,000 and $3,000, which is much lower than the cost for ASIC development tools. The FPGA based application development goes through a similar process as that of ASIC such as synthesis, functional and architectural simulation, timing analysis, place-and-route, etc. However, due to the reconfigurability of FPGA devices, their design insertion and de-

1

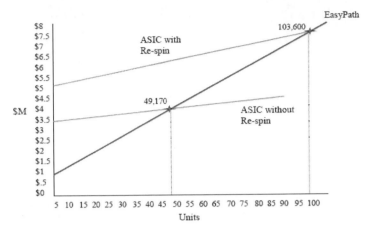

FIGURE 1.1:   Non-recurring engineering costs of EasyPath

sign verification are much simpler than ASICs. For example, users can use hardware co-simulation to simulate and verify the designs under test. Due to the relatively simpler development process, there are usually adequate tools provided by FPGA vendors at a very low costs. Many value-added tools from EDA vendors are also available for prices ranging from $20,000-$30,000.

• *No non-recurring engineering costs*

Non-recurring engineering (NRE) costs refer to the costs for creating a new product paid up front during a development cycle. The NRE costs are in contrast with "production costs", which are ongoing and are based on the quantity of material produced. For example, in the semiconductor industry, the NRE costs are the costs for developing circuit designs and photo-masks while the production costs are the costs for manufacturing each wafer.

Reconfigurable hardware is pre-fabricated and pre-tested before shipping to the customers. Once the final designs are complete, end users just need to download the bitstreams of the final designs to the FPGA devices. No photomasks are required. Therefore, there are no NRE costs for development using FPGAs. This would greatly reduce the overall costs and risks for application development, which is crucial considering the current economic downturns.

FPGA vendors provide an option for quickly converting the prototypes in FPGAs into structure ASIC chips or other low-cost alternatives with no or little changes to the original FPGA bitstreams. Examples include the structure ASIC based HardCopy technology [4] from Altera and the EasyPath technology [98] from Xilinx. Taking EasyPath as an example, there is no change to the FPGA bitstream during the conversion. The silicon used for the EasyPath device is identical to the FPGA silicon, which guarantees the identical functionalities between the two silicon devices. The EasyPath methodology uses a unique testing procedure to isolate the die that can be programmed in the factory with the customer's specific pattern. The customer receives Xilinx devices that are custom programmed and marked for their specific applications. The delivery time for these low-cost FPGA devices is a fraction compared with that of traditional ASIC devices. These customized structure ASIC chips are guaranteed to have the same functionalities as the original FPGA devices. No changes are introduced into the design at all. Finally, there is no tape-out or prototype cycle change and no change to the physical device itself. The NRE costs of the EasyPath technology compared with traditional ASIC design technologies are shown in Figure 1.1. We can see that for a large range of volume productions, the ASIC conversion option offered by FPGA vendors can have smaller NRE costs than traditional ASIC designs.

- *Fast time to market*

FPGAs require a negligible re-spinning time and offer a fast time-to-market. Once the design of the system is completed and verified, the whole product is ready for shipping to the customers. This is in contrast with ASIC based designs, which would incur up to a few months of turn-around time for producing the photo-masks.

- *Great product flexibility*

Various functionalities can be implemented on reconfigurable hardware through on-chip configurable logic, programmable interconnect, and heterogeneous pre-compiled hardware components available on modern FPGA devices. Reconfigurable hardware devices are widely used to implement various embedded systems. Recently, there have been active academic and industrial attempts to use reconfigurable hardware as accelerators for traditional CPUs. Scrofano et al. [81] evaluated the efficiency of developing floating-point high performance computing applications such as molecular dynamic simulation on FPGA devices.

With the aforementioned advantages, FPGAs have evolved into reconfigurable System-on-Chip (SoC) devices and are an attractive choice for implementing a wide range of applications. FPGAs are on the verge of revolutionizing many traditional computation application areas, which include digital signal processing, video and image processing, and high-performance scientific computing, etc.

1.2 Challenges and Contributions

The integration of multi-million-gate configurable logic, pre-compiled heterogeneous hardware components, and on-chip processor subsystems offer high computation capability and exceptional design flexibility to modern reconfigurable hardware. On the other hand, rapid energy estimation and energy efficient application synthesis using these hardware devices remain challenging research topics. Energy dissipation and efficiency have become a hurdle that prevents the further widespread use of FPGA devices in embedded systems, where energy efficiency is a key performance metric. The four major challenges for developing energy efficient applications using FPGAs are described in the following paragraphs.

- *The ever increasing design complexity makes low-level design flows and techniques unattractive for the development of many complicated systems.*

With FPGAs being used in many complicated systems, describing these systems using traditional register-transfer and gate-level ("low-level") design flow can turn out to be time consuming and unattractive in many design cases. Synthesis and placing-and-routing of a low-level design typically takes up to a couple of hours for a design consisting of multi-millions of logic gates. Even with technologies such as incremental synthesis and module-based low-level implementation, each minor change to the system could incur a significant amount of time to re-generate the low-level implementations. The time-consuming process of generating low-level implementations on FPGA devices prohibits efficient design space exploration and performance optimization.

When we consider the development of digital signal processing and MVI (multi-media, video and image) processing systems, the traditional low-level design flows mentioned above can significantly hamper the communication between hardware designers and algorithm developers. For instance, people from the signal processing community are usually not familiar with hardware description languages, while it is very demanding for a hardware designer to have a profound knowledge of various complicated digital signal processing and image processing algorithms. It is highly desired that domain-specific design tools be developed to bridge the communication gap between the hardware designers and algorithm developers.

- *State-of-the-art design and simulation techniques for general-purpose processors are inefficient for exploring the unprecedented hardware-software design flexibilities offered by reconfigurable hardware.*

From the design description perspective, register-transfer level and gate level techniques are inefficient for constructing reconfigurable platforms containing both hardware and software components. A true system-level design tool is required to describe the bus interfaces and connections between the hardware and software components and to associate software drivers with the hardware peripherals. Xilinx offers Platform Studio [97] for quickly constructing embedded processor systems on reconfigurable systems. Similarly, Altera offers SOPC (System on Programmable Chip) Builder [3] for their Nios processor systems. However, to develop customized hardware peripherals, users still need to reply on the traditional HDL based design flows. As we show later in this book, the RTL level simulation of a processor subsystems is a very time-consuming process and thus unattractive for practical software development.

From the simulation perspective, software programs usually execute for millions of clock cycles, which is far beyond the capabilities of even the fastest industrial RTL (Register Transfer Level) level simulators. The traditional approach for simulating the execution of software programs running on processors is through instruction-set simulators. Many academic and industrial instruction set simulators have been developed for various types of general-purpose processors. Examples of these instruction set simulators include SimpleScaler [14] and Amulator for StrongARM processors [7]. These simulators assume that the targeted processors have a relatively "fixed" hardware architecture. Some configuration options, such as cache size and associativities, memory sizes, etc., are usually available for these simulators. However, as is analyzed in Chapter 5, based on the assumption of relative "fixed" processor architectures, instruction-level simulators are unable to simulate the customized instructions and hardware peripherals attached to the processors. Hence, they are not suitable for hardware-software co-design.

- *Rapid energy estimation and energy performance optimization are challenging for systems using reconfigurable hardware.*

While energy estimation using low-level simulation can be accurate, it is time consuming and can be overwhelming considering the fact that there are usually many possible implementations of an application on FPGAs. Especially, the low-level simulation based energy estimation techniques are impractical for estimating the energy dissipation of the on-chip pre-compiled and soft processors. As we show in Chapter 5, simulating ~2.78 msec wall-clock execution time of a software program using post place-and-route simulation model of a state-of-the-art soft processor would take around 3 hours. The total time for estimating the energy dissipation of the software program is around 4 hours. Such an energy estimation speed prohibits the application of such low-level technique for software development considering the fact that many software programs are expected to run for tens and thousands of seconds.

On the other hand, the basic elements of FPGAs are look-up tables (LUTs), which are too low-level an entity to be considered for high level modeling and rapid energy estimation. It is not possible for a single high level model to capture the energy dissipation behavior of all possible implementations on FPGA devices. Using several common signal processing operations, Choi et al. [17] show that different algorithms and architectures used for implementing these common operations on reconfigurable hardware would result in significantly different energy dissipations. With this observation, a domain-specific energy performance modeling technique is proposed in [17]. However, to our best knowledge, there is no system-level development framework that integrates this modeling technique for energy efficient designs.

Considering the challenges discussed above, this manuscript makes the following four major contributions toward energy efficient application synthesis using reconfigurable hardware.

- *A framework for high-level hardware-software application development*

Various high-level abstractions for describing hardware and software platforms are mixed and integrated into a single and consistent application development framework. Using the proposed high-level framework, end users can quickly construct the complete systems consisting of both hardware and software components. Most importantly, the proposed framework supports co-simulation and co-debugging of the high-level description of the systems. By utilizing these co-simulation and co-debugging capabilities, the functionalities of the complete systems can be quickly verified and debugged without involving their corresponding low-level implementations.

- *Energy performance modeling for reconfigurable system-on-chip devices and energy efficient mapping for a class of application*

An energy performance modeling technique is proposed to capture the energy dissipation behavior of both the reconfigurable hardware platform and the target application. Especially, the communication costs and the reconfiguration costs that are pertinent to designs using reconfigurable hardware are accounted for by the energy performance models. Based on the energy models for the hardware platform and the application, a dynamic programming based algorithm is proposed to optimize the energy performance of the application running on the reconfigurable hardware platform.

- *A two-step rapid energy estimation technique and high-level design space exploration*

Based on the high-level hardware-software co-simulation framework proposed in the above section, we present a two-step rapid energy estimation technique. Utilizing the high-level simulation results, we employ an instruction-level energy estimation technique and a domain-specific modeling technique to provide rapid and fairly accurate energy estimation for hardware-software co-designs using reconfigurable hardware. Our technique for rapid energy estimation enables high-level design space exploration. Application developers can explore and optimize the energy performance of their designs early in the design cycle, before actually generating the low-level implementations on reconfigurable hardware.

- *Hardware-software co-design for energy efficient implementations of operating systems*

By utilizing the "configurability" of soft processors, we delegate some management functionalities of the operating systems to a set of hardware peripherals tightly coupled with the soft processor. These hardware peripherals manage the operating states of the various hardware components (including the soft processor itself) in a cooperative manner with the soft processor. Unused hardware components are turned off at the run-time to reduce energy dissipation. Example designs and illustrative examples are provided to show that a significant amount of energy can be saved using the proposed hardware-software co-design technique.

1.3   Manuscript Organization

Our contributions in the various aspects leading to energy efficient application synthesis using FPGAs are discussed in separate chapters of this book. Chapter 2 discusses the background and up-to-date information of reconfigurable hardware. Chapter 3 presents the framework for high-level hardware-software development using FPGAs. The development of reconfigurable hardware platforms that include multiple general-purpose processors using the proposed development framework is also discussed in this chapter. An energy performance modeling technique for reconfigurable hardware platforms that integrates heterogeneous hardware components as well as a dynamic-programming based performance optimization technique based on the energy models are proposed in Chapter 4. Based on the work in Chapter 3 and a domain-specific energy modeling performance modeling technique, a two-step rapid energy estimation technique and a design space exploration framework are discussed in Chapter 5. Chapter 6 puts forward a hardware-software co-design for energy efficient implementations of operating systems on FPGAs. Finally, we conclude in Chapter 7. The future research directions are also discussed in this chapter.

Chapter 2

Reconfigurable Hardware

2.1 Reconfigurable System-on-Chips (SoCs)

Reconfigurable hardware has evolved to become *reconfigurable system-on-chip devices*, which integrate multi-million gates of configuration logic, various heterogeneous hardware components, and processor cores into a single chip. The architectures of the configurable resources and the embedded hardware components are discussed in the following sections.

2.1.1 Field-Programmable Gate Arrays

Field-programmable gate arrays (FPGAs) are defined as programmable devices containing repeated fields of small logic blocks and elements. They are the major components of reconfigurable SoC devices. VLSI (Very Large Scale Integrated Circuits) devices can be classified as shown in Figure 2.1. FPGAs are a member of the class of devices called Field-Programmable Logic (FPL). It can be argued that FPGAs are an ASIC (Application-Specific Integrated Circuit) technology. However, the traditionally defined ASICs require additional semiconductor processing steps beyond those required by FPGAs. These additional steps provide performance advantages over FPGAs while introducing high non-recurring engineering (NRE) costs. Gate arrays, on the other hand, typically consist of a "sea of NAND gates" whose functions are provided by the customers in a "wire list". The wire list is used during the fabrication process to achieve the distinct definition of the final metal layer. The designer of an FPGA solution, however, has full control over the actual design implementation without the need (and delay) for any physical IC (Integrated Circuit) fabrication facilities.

2.1.1.1 Classifications

Field-programmable logics are available in virtually all memory technologies: SRAM, EPROM, E^{2}PROM, and anti-fuse. The specific memory technology determines whether a field-programmable logic device is *re-programmable* or *one-time programmable*, as well as the way it can be programmed. The configuration bitstreams of most modern field-programmable logic devices sup-

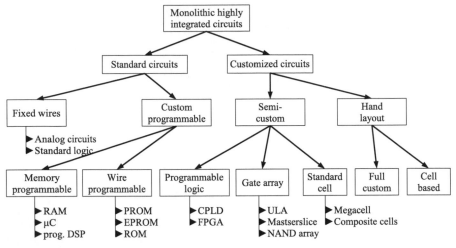

FIGURE 2.1:   Classification of VLSI devices

port both single-bit bitstreams and multi-bit streams. Compared with multi-bit streams, single-bit streams reduce the wiring requirements. But they also increase the programming time (typically in the range of ms). End users should choose the format of the bitstreams based on the start-up time and reconfiguration requirements of their target applications.

SRAM devices, the dominate technology of FPGAs, are based on static CMOS memory technology and are re-programmable and in-system programmable. The greatest advantages of SRAM-based field programmable devices are that they usually offer a higher computation capability and a better energy efficiency than reconfigurable devices built with other memory technologies. One major drawback of SRAM-based FPGA devices is that they usually require an external "boot" device for storing the configuration bitstreams. This has been changed recently with the release of Xilinx Spartan-3AN devices [107]. The Spartan-3AN devices have in-system flash memory for configuration and nonvolatile data storage, which eliminate the needs for external storage devices.

Electrically programmable read-only memory (EPROM) devices are usually used in a one-time CMOS programmable mode because of the need to use ultraviolet light for erasing the existing configuration. CMOS electrically erasable programmable read-only memory (E^2PROM) have the advantage of a short set-up time. Because the programming information is not "downloaded" to the device, it is better protected against unauthorized use. A recent innovation in the memory technologies for building reconfigurable hardware is called "Flash" memory and is based on the EPROM technology. These "Flash" based reconfigurable devices are usually viewed as "page-wise" in-system reprogrammable devices with physically smaller cells equivalent to that of E^2PROM devices.

TABLE 2.1: Various FPL technologies

Technology	SRAM	EPROM	E²PROM	Anti-fuse	Flash
Re-programmable	√	√	√	–	√
In-system programmable	√	–	√	–	√
Volatile	√	–	–	–	–
Copy protected	–	√	√	√	√
Example devices	Virtex 4/5	MAX5K	MACH	ACT	XC9500
	Spartan-3	Altera	AMD	Actel	Xilinx

Finally, the pros and cons of reconfigurable hardware based on different memory technologies are summarized in Table 2.1.

2.1.1.2 Configurable Logic Blocks

For SRAM (static random-access memory) based FPGA devices, the basic units are configurable logic blocks. These configurable logic blocks usually have a similar architecture. They contain one or more look-up tables (LUTs), various types of multiplexers, flip-flop registers as well as other hardware components. A k-input LUT requires 2^k SRAM cells and a 2^k-input multiplexer. A k-input LUT can implement any function of k-inputs. Previous research has shown that LUTs with 4-inputs lead to FPGAs with the highest area-efficiency [12]. Thus, for most commercial FPGAs, their LUTs accept four one-bit inputs and thus can implement the true table of any functions with four inputs. As FPGAs have a highly increased configurable logic density, devices with 6-input LUT have drawn significant industrial attentions. The ExpressFabric architecture of Xilinx Virtex-5 FPGA devices [109] uses 6-input LUTs for fewer logic levels and ultra-fast diagonal routing for reduced net delays. This has led to two speed-grade performance increases, 35% lower dynamic power reduction, and 45% less area compared with the previous 4-input LUT based Virtex-4 devices.

To reduce the time for synthesis, place-and-routing, the configurable logic blocks are usually evenly distributed through the device. Accordingly, there is a regular pattern for the programmable interconnection network for connecting these configurable logic blocks (see Section 2.1.1.3 for more information about the interconnection network).

As modern FPGAs contain up to multi-million gates of configurable logic, the configurable logic blocks are organized in a hierarchical architecture. Taking the Xilinx Virtex series FPGAs as an example, each CLB contains two slices and each slice contains four four-input LUTs. There are local interconnects between the two LUTs within a slice and between the two slices within a CLB. The arrangement of the slices within a CLB is shown in Figure 2.2.

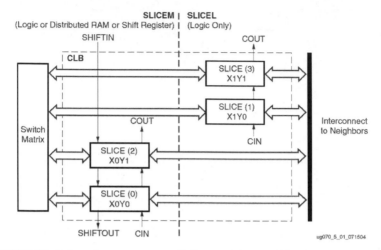

FIGURE 2.2:   Arrangement of slices within a Xilinx Virtex-4 CLB

TABLE 2.2:   Logic resources in a CLB

Slices	LUTs	Flip-flops	MULT ANDs	Arithmetic Carry-Chains	Distributed RAM	Shift Registers
4	8	8	8	2	64 bit	64 bit

In the most recent Virtex-4 FPGA devices, as shown in Table 2.2, each slice contains various hardware resources. The architectures and functionalities of these various hardware resources within a configurable logic block are discussed in detail in the following.

• *Look-up tables*: Virtex-4 function generators are implemented as 4-input look-up tables (LUTs). There are four independent inputs for each of the two function generators in a slice (F and G). The function generators are capable of implementing any arbitrarily defined four-input Boolean function. The propagation delay through a LUT is independent of the function implemented. Signals from the function generators can exit the slice (through the X or Y output), enter the XOR dedicated gate, enter the select line of the carry-logic multiplexer, feed the D input of the storage element, or go to the MUXF5.

As shown in Figure 2.3, the two 16-bit look-up tables within a slice can also be configured as a 16-bit shift register, or a 16-bit slice-based RAM. As is discussed in Section 2.1.1.3, dedicated routing resources are available among the slices for cascading and extending these shift registers and slice-based RAMs. Therefore, shift registers and slice-based RAMs of various lengths and dimensions can be configured on FPGA devices depending on the application requirements.

• *Multiplexers*: In addition to the basic LUTs, the Virtex-4 slices contain various types of multiplexers (MUXF5 and MUXFX). These multiplexers are

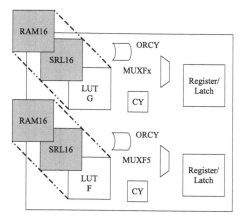

FIGURE 2.3: Virtex-4 slices within a Configurable Logic Block

used to combine two LUTs so that any function of five, six, seven, or eight inputs can be implemented in a CLB. The MUXFX is either MUXF6, MUXF7, or MUXF8 according to the position of the slice in the CLB. The MUXFX can also be used to map any function of six, seven, or eight inputs and selected wide logic functions. Functions with up to nine inputs (MUXF5 multiplexer) can be implemented in one slice. Wide function multiplexers can effectively combine LUTs within the same CLB or across different CLBs making logic functions with even more input variables.

• *Flip-flop registers and latches*: Each slice has two flip-flop registers/latches. The registers are useful for development of pipelined designs.

2.1.1.3   Routing Resources

There are three major kinds of routing architectures for commercial FPGA devices. The FPGAs manufactured by Xilinx, Lucent and Vantis employ an *island-style* routing architecture; the FPGAs manufactured by Actel employ a *row-based* routing architecture; and the FPGAs of Altera use a *hierarchical* routing architecture. In the following, we focus on the island-style routing architecture to illustrate the routing resources available on modern FPGA devices.

The routing architecture of an *island-style* FPGA device is shown in Figure 2.4. Logic blocks are surrounded by routing channels of pre-fabricated wiring segments on all four sides. The input or output pin of a logic block connect to some or all of the wiring segments in the channel adjacent to it via a *connection block* of programmable switches. At every intersection of a horizontal channel and a vertical channel, there is a *switch block*. The switch block is a set of programmable switches that allow some of the wire segments incident to the switch block to be connected to other wire segments. By turning on the appropriate switches, short wire segments can be connected

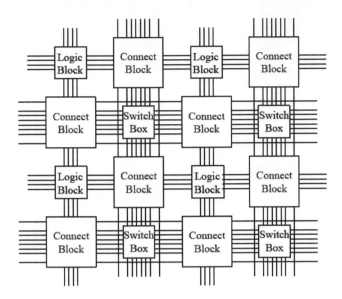

FIGURE 2.4: An FPGA device with island-style routing architecture

together to form longer connections. Note that there are some wire segments which remain unbroken through multiple switch blocks. These longer wires span multiple configurable logic blocks.

We will take a close look at the popular Xilinx Virtex-II Pro devices to illustrate the modern island-style routing architecture. Virtex-II Pro have fully buffered programmable interconnections, with a number of resources counted between any two adjacent switch matrix rows or columns. With the buffered interconnections, fanout has minimal impact on the timing performance of each net.

There is a hierarchy of routing resources for connecting the configurable logic blocks as well as other embedded hardware components. The global and local routing resources are described as follows.

- *Long lines*: the long lines are bidirectional wires that distribute signals across the device. Vertical and horizontal long lines span the full height and width of the device.

- *Hex lines*: the hex lines route signals to every third or sixth block in all four directions. Organized in a staggered pattern, hex lines can only be from one end. Hex-line signals can be accessed either at the endpoints or at the middle point (i.e. three blocks from the source).

- *Double lines*: the double lines route signals to every first or second block in all four directions. Organized in a staggered pattern, double lines can be driven only at their endpoints. Double-line signals can be accessed either at the endpoints or at the midpoint (one block away from the source).

- *Direct connect lines* the direct connect lines route signals to neighboring

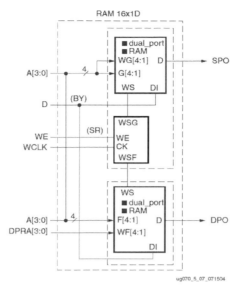

FIGURE 2.5: Dual-port distributed RAM (RAM16x1D)

blocks: vertically, horizontally, and diagonally.

• *Fast connect lines*: the fast connect lines are the internal CLB local interconnects from LUT outputs to LUT inputs. As shown in Figure 2.5, a 16x1D dual-port distributed RAM can be constructed using the fast connect lines with a CLB.

In addition to the global and local routing resources, dedicated routing signals are available.

• There are eight global clock nets per quadrant.

• Horizontal routing resources are provided for on-chip tri-state buses. Four partitionnable bus lines are provided per CLB row, permitting multiple buses within a row.

• Two dedicated carry-chain resources per slice column (two per CLB column) propagate carry-chain MUXCY output signals vertically to the adjacent slice.

• One dedicated SOP (Sum of Product) chain per slice row (two per CLB row) propagate ORCY output logic signals horizontally to the adjacent slice.

• As shown in Figure 2.6, there are dedicated routing wires, which can be used to cascadable shift register chains. These dedicated routing wires connect the output of LUTs in shift register mode to the input of the next LUT in shift-register mode (vertically) inside the CLB.

The routing architecture may be different as the density of modern FPGA devices continues increasing. An improved reduced-hop architecture is introduced in Xilinx Virtex-5 devices, which are built based on 6-input look-up tables. For FPGAs with 6-input LUTs, the routing requirements between

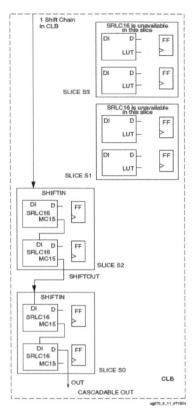

FIGURE 2.6:   Cascadable shift registers

near-by configurable logic blocks have become more challenging. Therefore, instead of using the asymmetric routing architecture described in the previous paragraphs, the Virtex-5 devices have diagonally symmetric interconnects between adjacent configurable logic blocks, which is shown in Figure 2.7.

2.1.2   Pre-Compiled Embedded Hardware Components

It is becoming popular to integrate pre-compiled embedded hardware components into a single FPGA device. The embedded hard and soft processor cores enable software development on reconfigure hardware platform.

2.1.2.1   Embedded General-Purpose Processors

As shown in Figure 2.8, the PowerPC 405 processor embedded in Virtex-4 consists of two major components: the PowerPC 405 core and the auxiliary functional unit (APU).

- *Pre-Compiled Processor Cores*

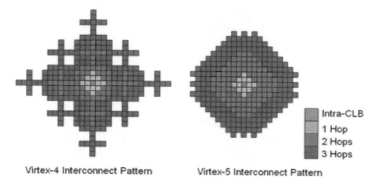

Virtex-4 Interconnect Pattern   Virtex-5 Interconnect Pattern

FIGURE 2.7: Diagonally symmetric interconnects in Virtex-5

The PowerPC 405 core is responsible for the execution of embedded software programs. The core contains four major components: central processing unit (CPU), memory management unit (MMU), cache unit, and timer and debug unit. The PowerPC 405 CPU implements a 5-stage instruction pipeline, which consists of fetch, decode, execute, write-back, and load write-back stages. The PowerPC 405 cache unit implements separate instruction-cache and data-cache arrays. Each of the cache arrays is 16 KB in size, two-way set-associative with 8 word (32 byte) cache lines. Both of the cache arrays are non-blocking, allowing the PowerPC 405 processor to overlap instruction execution with reads over the PLB (processor local bus) when cache misses occur.

• *Instruction Extension*

The architecture of the APU controller is illustrated in Figure 2.8. The FCM (Fabric Co-processing Module) interface is a Xilinx adaptation of the native Auxiliary Processor Unit interface implemented on the IBM processor core. The hard core APU Controller bridges the PowerPC 405 APU interface and the external FCM interface.

Using the APU controller, the user can extend the instruction set of the embedded PowerPC 405 processor through a tight integration of the processor with the surrounding configurable logic and interconnect resources. The actual execution of the extended instructions is performed by the FCM. There are two types of instructions that can be extended using an FCM, that is, pre-defined instructions and user-defined instructions. A *pre-defined instruction* has its format defined by the PowerPC instruction set. A floating point unit (FPU) which communicates with the PowerPC 405 processor core through the FCM interface is one example of the APU instruction decoding. While the embedded PowerPC 405 only supports fixed-point computation instructions, floating-point related instruction are also defined in the PowerPC 405 instruction set. The APU controller handles the decoding of all PowerPC floating-point related instructions, while the FCM is responsible for calculating the floating-point results. In contrast, a *user-defined instruction* has a

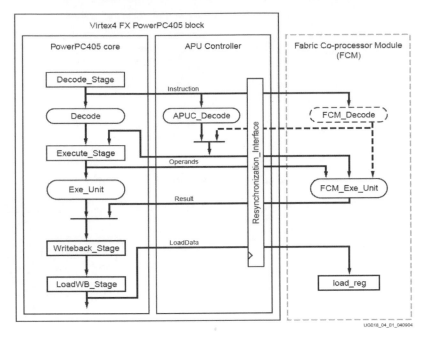

FIGURE 2.8:   Pipeline flow architecture of the PowerPC 405 Auxiliary Processing Unit

configurable format and is a much more flexible way of extending the PowerPC instruction set architecture (ISA). The decoding of the two types of FCM instruction discussed above can be done either by the APU controller or by the FCM. In case of *APU controller decoding*, the APU controller determines the CPU resources needed for instruction execution and passes this information to the CPU. For example, the APU controller will determine if an instruction is a load, a store, or if it needs source data from the GPR, etc. In case of *FCM instruction decoding*, the FCM can also perform this part of decoding and pass the needed information to the APU controller. In case that an FCM needs to determine the complete function of the instruction, the FCM must perform a full instruction decode as well as handle all instruction execution.

The APU controller also performs clock domain synchronization between the PowerPC clock and the relatively slow clocks of the FCM. The FCMs are realized using the configurable logic and routing resources available on the FPGA device as well as other pre-compiled hardware resources. Limited by the configurable hardware resources, the FCMs usually have slower maximum operating frequencies than the pre-compiled embedded general-purpose processor. In many practical application developments, the embedded processor and the FCM are operating at different clock frequencies. In this case, the APU controller synchronizes the operations in different clock domains.

• *Bus hierarchy* The embedded PowerPC 405 processor supports the IBM

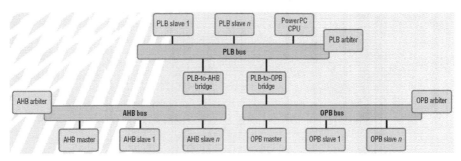

FIGURE 2.9: IBM CoreConnect bus architecture

CoreConnect bus suite [49], which is shown in Figure 2.9. This include PLB (Processor Local Bus) buses for connecting high-speed, high-performance hardware peripherals, OPB (On-chip Peripheral Bus) buses for connecting low-speed, low-performance hardware peripherals (*e.g.*, the UART serial communication hardware peripheral) and DCR (Device Control Register) buses for queuing the status of the attached hardware peripherals.

2.1.2.2 Pre-Compiled Memory Blocks

Modern programmable devices integrate up to 10 Mbits of pre-compiled dedicated memory blocks. The Xilinx Virtex-II Pro devices integrate up to 444 block RAM (BRAM) blocks. Each of the BRAM blocks holds 18 Kbits of data. This amounts to a data storage up to 7,992 Kbits. The Xilinx Virtex-5 devices integrate up to 244 BRAM blocks. Each of the BRAM blocks holds 36 Kbit of data and can be configured as two independent 18-Kbit memory blocks. This amounts to a data storage up to 8,784 Kbits. Depending on the specific application requirements, these pre-compiled memory blocks can be configured as different widths and word-lengths.

These pre-compiled memory blocks are an important hardware resource for embedded software development. The memory blocks can be configured as cache and/or on-chip memory systems for the embedded hard/soft processor cores. From the energy efficient perspective, the research work by Choi, et al. [17] shows that these pre-compiled memory blocks can lead to significant energy savings as storage for moderate to large amounts of data.

2.1.2.3 Pre-Compiled Digital Signal Processing Blocks

Embedded DSP blocks are available on most modern field programmable devices for performing computation intensive arithmetic operations, which are required by many digital and image signal processing applications. These DSP blocks are buried among the configurable logic blocks and are usually uniformly distributed throughout the devices. The Xilinx Virtex-II series and Spartan-3 series FPGAs offer up to 556 dedicated multipliers on a single chip. Each of these embedded multipliers can finish signed 18-bit-by-18-bit multiplication in one clock cycle at an operating frequency up to 350 MHz. The Xilinx Virtex-4/Virtex-5 devices have various versions of DSP48 blocks. Each of these blocks can perform multiplication, addition/subtraction, and other operations in a single clock cycle with an operating frequency up to 550 MHz. The integration of pre-compiled DSP block would provide modern FPGA devices with an exceptional computation capability.

One architecture of the DSP48 blocks on the Xilinx Virtex-4 devices is shown in Figure 2.10. These DSP48 blocks can be configured to efficiently perform a wide range of basic mathematical functions. This includes adders, subtracters, accumulators, MACCs (Multiply-and-Accumulates), multiply multiplexers, counters, dividers, square-root functions, shifters, etc. The DSP48 blocks stack vertically in a DSP48 column. The height of a DSP48 slice is the same as four CLBs and matches the height of one BRAM block. This regularity alleviates the routing problem resulting from the wide data paths of the DSP48 blocks.

There are optional pipeline stages within the DSP48 blocks in order to ensure high performance arithmetic functions. The input and output of the DSP48 blocks can be stored in the dedicated registers within the tiles. Besides, the DSP48 blocks are organized in columns. There are dedicated routing resources among the DSP48 blocks within a column. These dedicated routing resources provide fast routing between DSP48 blocks, while significantly reducing the routing congestions to the FPGA fabric. The tiled architecture allows for efficient cascading of multiple DSP48 slices to form various mathematical functions, DSP filters, and complex arithmetic functions without the use of general FPGA fabric and routing resources. Designs using these pre-compiled DSP blocks can achieve significant power reduction, while delivering very high performance and efficient silicon utilization.

Several variances of the DSP48 blocks are availabe on the Xilinx devices in order to better perform various DSP algorithms. An 18-bit pre-adder is added to the DSP48A blocks in Spartan-3A DSP devices. The hard pre-adder significantly increases the density, performance, and power efficiency of symmetric FIR filters and complex multipliers.

2.1.3   Soft Processors

2.1.3.1   Definition and Overview

FPGA configured soft processors are RISC (Reduced Instruction-Set Computer) processors realized using configurable resources available on FPGA devices. They have become very popular in recent years and have been adopted in many customer designs. Examples of 32-bit soft processors include Nios from Altera [3], ARM7-based CoreMP7 from Actel [1], OpenRISC from Open-Cores [55], SPARC architecture based LEON3 from Gaisler Research [30], and MicroBlaze from Xilinx [97]. For these 32-bit soft processors, a complete C/C++ development tool chain is usually provided by their vendors and de-

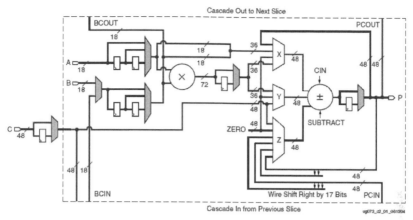

FIGURE 2.10: Architecture of DSP48 slices on Xilinx Virtex-4 devices

velopers. Xilinx provides an integrated Embedded Development Kit (EDK) to facilitate the development on their MicroBlaze processor.

There are also several popular 8-bit micro-controller type soft processors available, such as the Xilinx 8-bit PicoBlaze [100] and PacoBlaze [54], which is the open-source soft processor compatible with the PicoBlaze processor.

• *Three-Dimensional Design Trade-Offs*

The major application development advantage of using soft processors is that they provide new design trade-offs by time sharing the limited hardware resources (e.g., configurable logic blocks on Xilinx FPGAs) available on the devices. Many management functionalities and computations with tightly coupled data dependency between calculation steps (e.g., many recursive algorithms such as the Levinson Durbin algorithm [42]) are inherently more suitable for software implementations on processors than the corresponding customized (parallel) hardware implementations. Their software implementations are more compact and require a much smaller amount of hardware resources. In some design scenarios, compact designs using soft processors can effectively reduce the static energy dissipation of the complete system by fitting themselves into smaller FPGA devices [93].

Various embedded operating systems have been ported to run on soft processors. This includes full-fledged operating systems such as Linux, uCLinux (a derivative of Linux kernel intended for micro-controllers without memory management units), eCos, etc. There are also smaller foot-print real-time operating systems ported to soft processors, such as MicroC/OS-II, PrKernel (μITRON4.0), ThreadX, Nucleus, etc. These operating systems running on soft processors significantly enhance the ability of reconfigurable hardware to communicate with various interfaces (e.g., Ethernet, USB), and expand the application of reconfigurable hardware in embedded system development.

• *"Configurability"*

Since soft processors are realized using reconfigurable hardware, they can be "configurable", either at the compile time or at the run-time, and thus enable the customization of the instruction set and/or the attachment of customized hardware peripherals. The Nios processor allows users to customize up to five instructions. The MicroBlaze processor supports various dedicated communication interfaces for attaching customized hardware peripherals to it. Such customization can lead to computation speed-up for algorithms with a large degree of parallelism. It can also lead to efficient execution of some auxiliary management functionalities, which are described in detail in Chapter 6.

Many soft processors support a hierarchy of shared buses for connecting peripherals with different communication requirements. For the CoreMP7 soft processor as shown in Figure 2.11, it supports both the AHB (advanced high-performance bus) and APB (advanced peripheral bus) interfaces.

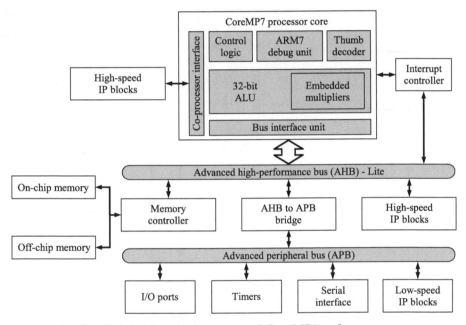

FIGURE 2.11:   Architecture of CoreMP7 soft processor

- *On-Chip Multi-Processor Development*

Configurable multi-processor platforms are implemented by instantiating multiple hard and/or soft processor cores, and connecting them through an application-specific communication network on a single reconfigurable hardware. There are various topologies between the multiple processors. James-Roxby et al. [50] implement a network of MicroBlaze processors with a star topology. A multi-processor platform with a two-dimensional topology is available from CrossBow Technologies [31].

There are several major advantages offered by configurable multi-processor

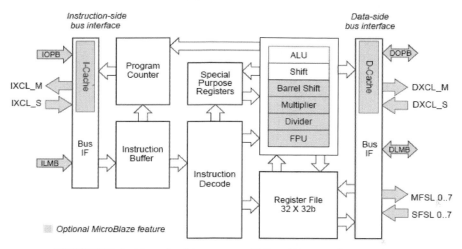

FIGURE 2.12: Architecture of MicroBlaze soft processor

platforms. The co-operation between multiple processors improves the processing and interfacing capabilities of reconfigurable hardware, compared with single-processor solutions. Users can reuse legacy software programs to meet their design goals, which greatly reduces the development cycle and time-to-market. There are many existing C programs for realizing various computation and management functionalities. There are also many C/C++ application specific interfaces (APIs) to coordinate the execution between multiple processors (e.g., the open-source message passing interface for parallel processing [28], and parallel virtual machines (PVM) [35]). These common software functionalities can be readily ported to configurable multi-processor platforms with little or no changes.

Another major advantage of configurable multi-processor platforms is that they are implemented using reconfigurable hardware, and thus are "configurable" and highly extensible by allowing the attachment of customized hardware peripherals. In addition to the traditional approaches of distributing the software execution among the multiple processors, the application designer can customize the hardware platform and optimize its performance for the applications running on them. Two major types of customization can be applied to configurable multi-processor platforms for performance optimization. (1) *Attachment of hardware accelerators*: as described in the previous paragraphs, the application designer can optimize the performance of a single soft processor by adding dedicated instructions and/or attaching customized hardware peripherals to it. These dedicated instructions and customized hardware peripherals are used as hardware accelerators to speed up the execution of the time-consuming portions of the target application (e.g., the computation of Fast Fourier Transform and Discrete Cosine Transform). The Nios processor allows users to customize up to five instructions. The MicroBlaze processor and the LEON3 processor support various dedicated interfaces and bus protocols for attaching customized hardware peripherals to them. The

LEON3 processor even allows the application designer to have fine control of the cache organization. Cong et al. show that adding customized instructions to the Nios processor using a shadow register technique results in an average 2.75x execution time speed-up for several data-intensive digital signal processing applications [20]. Also, by adding customized hardware peripherals, the FPGA configured VLIW (Very-Long Instruction Word) processor proposed by [52] achieves an average 12x speed-up for several digital signal processing applications from the MediaBench benchmark [57]. The customization of a single soft processor is further discussed in the section on background-customization. (2) *Application-specific efficient communication network and task management components among the multiple processors*: based on the application requirements, the end user can instantiate multiple soft processors, connect them with some specific topologies and communication protocols, and distribute the various computation tasks of the target application among these processors. James-Roxby et al. develop a configurable multi-processor platform for executing a JPEG2000 encoding application [21]. They construct the multi-processor platform by instantiating multiple MicroBlaze processors and connecting them with a bus topology. They then employ a Single-Program-Multiple-Data (SPMD) programming model to distribute the processing of the input image data among the multiple MicroBlaze processors. Their experimental results in [50] show that a significant speed-up is achieved for the JPEG2000 application on the multi-processor platform compared with the execution on a single processor. Jin et al. built a configurable multi-processor platform using multiple MicroBlaze processors arranged as multiple pipelined arrays for performing IPv4 packet forwarding [51]. Their multi-processor platform achieves 2x time performance improvement compared with a state-of-the-art network processor.

2.1.3.2   Customization

There are two major ways of customizing soft processors, which are discussed as follows.

- *Tight Coupling through Instruction Customization*

The instructions that are not needed by the target applications and the corresponding hardware components can be removed from the soft processors in order to save the hardware resources and improve their timing and energy performance. Considering the architecture of the MicroBlaze processor shown in Figure 2.12, the MicroBlaze processor can be configured without the floating-point unit (FPU) and the related instructions for many image processing applications. These applications usually only require fixed-point operations. Removing the floating-point unit and the related instructions can reduce more than 60% of the hardware resources required by MicroBlaze [97].

There are two primary approaches for extending the instruction set of soft processors. One approach is to share the register files between the soft processors and the customized hardware peripherals. One example is the MicroBlaze

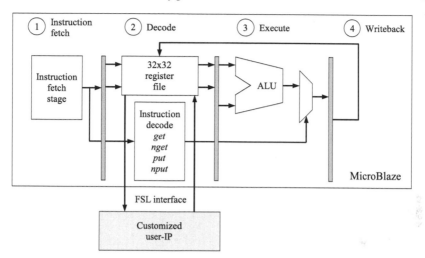

FIGURE 2.13: Extension of the MicroBlaze instruction set through FSL interfaces

processor as shown in Figure 2.13. MicroBlaze provides a set of reserved instructions and the corresponding bus interfaces that allow the attachment of customized hardware peripherals. The main advantage of the register-file based approach is easy development. The design of the attached hardware peripherals are independent of the soft processors and would not affect their performance for executing the embedded software programs as well as their interactions with other hardware peripherals.

As shown in Figure 2.14, the other approach of instruction extension is through connecting the customized hardware peripherals in parallel with the ALU (Arithmetic Logic Unit) of the processors. This direct ALU coupling approach is adopted by the Altera Nio soft processor. This coupling approach provides more design flexibility as the end users can be involved in the instruction decoding process and get access to various low-level functionalities of the processor core. It can lead to a potentially tighter integration between the hardware peripherals and the processor core compared with the register file based coupling approach. Note that extra attention is required during development to make sure that the attached hardware peripherals do not become the critical path of the processor and thus do not affect the maximum operating frequencies that can be achieved by the soft processor core. The end users are responsible for properly pipelining the hardware peripherals to ensure the overall timing performance of the soft processor.

• *Hierarchical Coupling through Shared Bus Interfaces*

Various customized hardware peripherals with different processing capabilities and data transmission requirements may be connected to the processors. Hierarchical coupling ensures efficient attachment of the various customized hardware peripherals to the processors by grouping them into differ-

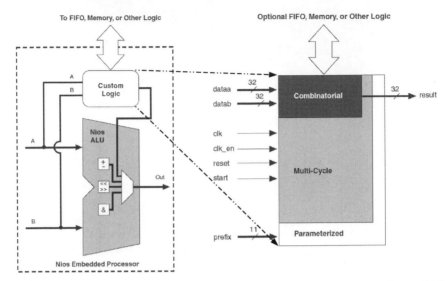

FIGURE 2.14:   Extension of the Nios instruction set through direct ALU coupling

ent hierarchies. The two *de facto* hierarchical bus architectures are discussed below.

- *CoreConnect Bus Architecture* includes the Processor Local Bus (PLB), the On-chip Peripheral Bus (OPB), a bus bridge, two arbiters, and a Device Control Register (DCR) bus [49]. It is an IBM proprietary on-chip bus communication link that enables chip designs from multiple sources to be interconnected to create new chips. It is widely adopted by the PowerPC 405 processors embedded in the Xilinx Virtex-II Pro and Virtex-4 FPGAs as well as the MicroBlaze soft processors. The PLB bus is used for tight integration of high performance hardware peripherals to the processor core while the OPB bus is used for connecting low-speed hardware peripherals to the processor core. Both the PLB and OPB buses form a star network for the hardware peripherals connecting them. In contrast, the DCR bus connects the hardware peripherals in a serial manner. Thus, its performance and response time degrade in proportion to the hardware peripherals connected to it. The DCR bus is mainly used for querying the status of the hardware peripherals.

- *Advanced Microprocessor Bus Architecture (AMBA)* is an open bus architecture standard maintained by ARM. Similar to the CoreConnect bus architecture, AMBA defines a multilevel busing system, with a system bus and a lower-level peripheral bus. These include two system buses: the AMBA High-Speed Bus (AHB) or the Advanced System Bus (ASB), and the Advanced Peripheral Bus (APB). The AHB bus can be used to connects embedded processors such as an ARM core to high-performance peripherals, DMA (Direct Memory Access) controllers, on-chip memory and interfaces. The APB is designed with a simpler bus protocol and is mainly used for ancillary or general purpose peripherals.

2.1.4 Domain-Specific Platform FPGAs

The integration of various pre-compiled hardware components has transformed reconfigurable hardware to become heterogeneous processing platforms. They are used in many application areas, ranging from wireless baseband digital signal processing to embedded processing in assorted consumer electronic devices. Digital signal processing applications usually require many efficient arithmetic computation blocks (e.g., the DSP blocks discussed in Session 2.1.2.3) for various filtering calculations. Embedded processing applications may need more dedicated memory blocks to store and cache the software programs running on the on-chip processor cores. Control applications may require a lot of configurable look-up tables to realize complicated control and management functionalities. These different applications pose vastly different development requirements to the underlying reconfigurable hardware. It is impossible to have a single heterogeneous reconfigurable hardware platform that is optimized across such a large range of applications.

To help alleviate these design issues, Xilinx has developed an FPGA architecture in recent years, coined ASMBL (Application Specific Modular Block Architecture) [44]. The ASMBL architecture enables rapid cost-effective deployment of multiple domain-specific FPGA platforms with a blend of configurable slices and pre-compiled hardware components optimized for the specific domain.

ASMBL supports the concept of multiple domain-specific platforms through the use of a unique column-based architecture, which is shown in Figure 2.15. Each column represents a silicon sub-system with specific capabilities, such as configurable logic cells, memory blocks, I/O interfaces, digital signal processing blocks, embedded processor cores, and mixed signal components. These domain-specific FPGAs are assembled by combining columns with different capabilities into a single chip targeting a particular class of applications. This is opposed to application specific platforms which would address a single application. The typical application domains include logic-intensive, memory-intensive, or processing-intensive. For example, a processing-intensive chip for graphics processing might have more columns devoted to DSP functions than would a chip targeting an application in the logic domain.

Using the ASMBL technology, the Virtex-5 family of FPGAs offers a choice of four different reconfigurable hardware platforms, which are shown in Figure 2.16. Each of the hardware platforms delivers an optimized balance of high-performance logic, serial connectivity, signal processing, and embedded processing. The Virtex-5 LX platforms are optimized for high-performance logic with the highest percentage of configurable slides and the least percentage of pre-compiled hardware components. The LXT platforms are optimized for low-power serial connectivity. They integrate pre-compiled PCI Express Endpoint blocks, up to 6 Ethernet MAC (media access control) blocks, and up to 24 RocketIO transceivers for gigabit serial communication. The SXT platforms are optimized for digital signal processing and memory-intensive

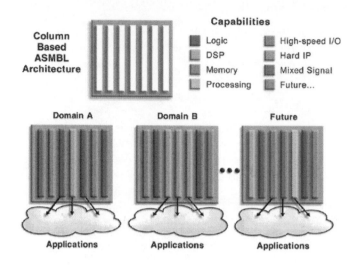

FIGURE 2.15:   Architecture of ASMBL domain-specific platform FPGAs

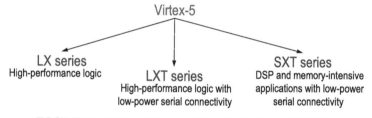

FIGURE 2.16:   Virtex-5 Multi-Platform FPGAs

applications low-power serial connectivity. In addition to the dedicated interfacing blocks in the LXT platforms, the SXT platforms contain the largest amount of DSP48E blocks for targeting digital signal processing and image processing application domains. Lastly, the FXT platforms are optimized for embedded processing applications. While details of the FXT platforms are not available at the time of writing, they are supposed to contain the largest amount of embedded memory blocks of all the Virtex-5 domain-specific FPGA devices.

Another example of domain-specific platform FPGAs are the Spartan-3 FPGA devices, which offer the following five different platforms.

- Spartan-3A DSP Platform: This platform contains dedicated DSP48A slices, which can perform efficient computations, including multiplication, multiply-accumulate (MACC), multiplier followed by an addition/subtraction, three-input addition, barrel shift, wide bus multiplexing, and magnitude compare. It is optimized for digital signal process-

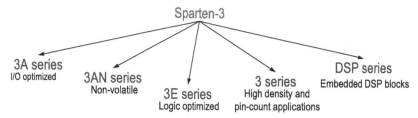

FIGURE 2.17: Spartan-3 Multi-Platform FPGAs

ing applications. It is suitable for applications that require DSP MACs (multiply-and-accumulates) and expanded memory, and for designs requiring low cost FPGAs for signal processing applications (e.g., military radio, surveillance cameras, medical imaging, etc.)

- Spartan-3AN Platform: This platform features an integrated 16-Mbit flash memory within a single device. It supports flexible power-management modes with over 40% power savings in suspend mode. The Spartan-3AN platform targets applications where non-volatile, system integration, security, and large user flash are required. It is designed for application domains, such as space-critical or secure applications as well as low cost embedded controllers.

- Spartan-3A Platform: This platform is optimized for I/O interfacing and has pre-compiled hardware components that support the widest kinds of I/O standards among the Spartan-3 FPGA series. It is intended for applications where I/O count and capabilities matter more than logic density. It is ideal for bridging, differential signaling, and memory interfacing applications. These target applications usually require wide or multiple interfaces and modest processing

- Spartan-3E Platform: This platform is logic optimized and contains the least percentage of hard IP blocks.

- Spartan-3 Platform This platform has the highest density and pin-count. It is for applications where both high logic density and high I/O count are important (e.g., highly-integrated data-processing applications).

2.2 Design Flows

On the one hand, the ever increasing logic density and the availability of heterogeneous hardware components offer exceptional computation capability and design flexibility to reconfigurable hardware. On the other hand, however, the complexity of developing applications on reconfigurable hardware has drastically increased, and has started to hamper the further adoption of reconfigure hardware. There have been efforts from both the industry and academia to drive the development of novel design tools in order to help ease the design difficulties.

We provide a summary of the state-of-the-art design flows in this section. The design flows of application development using reconfigurable hardware can be roughly classified into two categories: the traditional low-level register-transfer/gate level based design flows and high-level design flows. Our focus is on the various tools based on the high-level design flows. In particular, we provide an in-depth discussion of the Xilinx *System Generator for DSP* tool in the next section as the energy efficient application development techniques and their illustrative examples presented in this manuscript are mainly based on this design tool.

2.2.1   Low-Level Design Flows

The traditional low-level design flows for application development using reconfigurable hardware are shown in Figure 2.18. The application developer uses hardware description languages (HDLs) (such as VHDL or Verilog) to describe the low-level (i.e., register-transfer and/or gate-level) architecture and behavior of the target systems. Very often, in order to satisfy some specific application requirements (e.g., minimum operating frequencies and hardware resource budgets) and achieve higher performance, design constraints are provided to the low-level design tools along with the low-level design descriptions. There are two kinds of design constraints. One kind of design constraints is area constraints, which includes the absolute or relative locations of the input/output pins and the hardware bindings of some operations. For example, the developer can specify that the multiplication operations are realized using the embedded multiplier or DSP blocks, and the data is stored using embedded memory blocks, rather than the corresponding configurable logic slice based implementations. The other kind of design constraints is timing constraints. Most of the low-level designs provide either a graphical interface (e.g., the PlanAhead tool [105]) or a text file to report the critical paths of a design. The application developers can then add timing constraints and force the placing-and-routing tools to assign higher priorities on resolving the congested placement and routing issues on critical paths, and they can spend more efforts to meet the timing requirements of these critical paths.

Before synthesis and the generation of low-level implementations, the application developer can perform functional simulation to verify the functional behavior of the target systems, regardless of the actual hardware implementations and logic and transfer delays. After synthesizing and placing-and-routing the low-level designs, the application developer can perform architectural and timing simulation to further verify the correctness of their systems. Once the

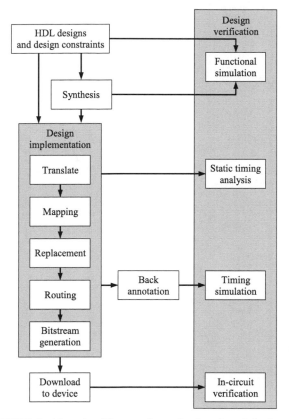

FIGURE 2.18: An illustration of a low-level design flow

functional, structural and timing correctness of the systems is verified, the end user can download the low-level bitstreams generated based on the low-level design descriptions and the design constraints onto reconfigurable hardware.

2.2.2 High-Level Design Flows

As modern reconfigurable hardware are being used to implement many complex systems, low-level design flows can be inefficient in many cases. High-level design flows based on high-level modeling environments (such as MAT-LAB/Simulink) and high-level languages (such as C/C++ and Java) are becoming popular for application development using reconfigurable hardware.

One important kind of high-level design tools are those based on MAT-LAB/Simulink. These include *System Generator* from Xilinx [108], *DSP Builder* from Altera [3], *Synplify DSP* from Synplify [5], and *HLD Coder* from MathWorks.

Another important kind of high-level design tools are those based on high-level language, such as C/C++, MATLAB M code, etc.

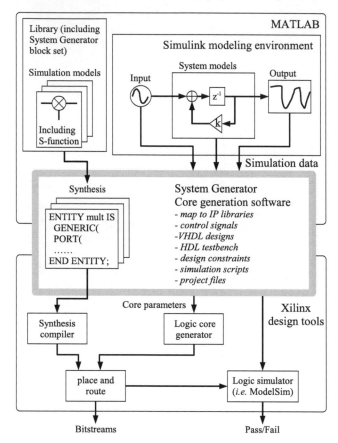

FIGURE 2.19: Design flow of *System Generator*

- *MATLAB M code based AccelDSP Synthesis Tool*

AccelDSP Synthesis Tool from Xilinx [104] is a high-level MATLAB language based tool for designing DSP blocks for Xilinx FPGAs. The tool automates floating-point to fixed-point conversion, generates synthesizable VHDL or Verilog from MATLAB language, and also creates testbenches for verification. It can also be used to generate fixed-point C++ simulation models or corresponding Simulink based System Generator block models from a MATLAB algorithm.

- *ImpulseC from Impulse Accelerated Technology, Inc*

ImpulseC extends standard C to support a modified form of the communicating sequential processes (CSPs) programming model. The CSP programming model is conceptually similar to the data-flow based ones. Compared with data-flow programming models, CSP based programming models put more focus on simplifying the expression of highly parallel algorithms through the use of well-defined data communication, such as message passing and synchronization mechanisms.

In ImpulseC applications, hardware and software elements are described in processes and communicate with each other primarily through buffered data streams that are implemented directly in hardware. The data buffering, which is implementing using FIFOs that are specified and configured by the ImpulseC programmer, makes it possible to write parallel applications at a relatively high level of abstraction. The ImpulseC application developers can be freed of clock-cycle by clock-cycle synchronization.

ImpulseC is mainly designed for dataflow-oriented applications, but is also flexible enough to support alternate programming models including the use of shared memory as a communication mechanism. The programming model that an ImpulseC programmer selects will depend on the requirements of the application and on the architectural constraints of the selected programmable platform target. ImpulseC is especially suitable for developing applications that have the following properties:

▷ The application features high data rates to and from different data sources and between processing elements.

▷ Data sizes are fixed (typically one byte to one word), with a relatively small stream payload to prevent processes from becoming blocked.

▷ Multiple related but independent computations are required to be performed on the same data stream.

▷ The data consists of low or fixed precision data values, typically fixed width integers or fixed point fractional values.

▷ There are references to local or shared memories, which may be used for storage of data arrays, coefficients and other constants, and to deposit results of computations.

▷ There are multiple independent processes communicating primarily through the data being passed, with occasional synchronization being requested via messages.

The ImpulseC library provides minimal extensions to the C language (in the form of new data types and intrinsic function calls) that allow multiple, parallel program segments to be described, interconnected and synchronized. The ImpulseC compiler translates and optimizes ImpulseC programs into appropriate lower-level representations, including RTL-level VHDL that can be synthesized to FPGAs and standard C (with associated library calls) that can be compiled onto supported microprocessors through the use of widely available C cross-compilers.

In addition to ImpulseC, there are other C based design tools, such as CatapultC from Mentor Graphics [62] and PICO Express from Synfora [89]. In contrast to ImpulseC, these design tools are based on standard C code, rather than providing a minimal extension. A special coding style of the C code is required to derive the desired low-level hardware implementations. The similarity with ImpulseC is that these C based design tools can greatly speed up the development process and simulation speed. This would facilitate design space exploration and design verification.

2.2.3   System Generator for DSP

More details of the Xilinx System Generator for DSP tool (referred to as "System Generator"in the following) is introduced in this section. The *System Generator* tool is the primary tool used throughout this manuscript to demonstrate the various rapid energy estimation and optimization techniques proposed by the authors.

2.2.3.1   High-Level Modeling and Design Description

The overall design flow of *System Generator* is illustrated in Figure 2.19. *System Generator* provides an additional set of blocks, accessible in the Simulink library browser, for describing hardware designs deployed onto Xilinx FPGAs. A screenshot of the Xilinx block set is shown in Figure 2.20. The *System Generator* blocks can interact and be co-simulated with other Simulink blocks. Only blocks and subsystems consisting of blocks from the Xilinx block set are translated by System Generator into hardware implementations. Each of the *System Generator* blocks is associated with a parameterization GUI (Graphical User Interface). These GUI interfaces allow the user to specify the high-level behaviors of the blocks, as well as low-level implementation options (e.g., target FPGA device, target system clock period). The automatic translation from the Simulink model into hardware realization is accomplished by mapping the Xilinx blocks into IP (Intellectual Property) library modules, inferring control signals and circuitry from system parameters (e.g., sample periods and data types), and finally converting the Simulink hierarchy into a hierarchical VHDL netlist. In addition, *System Generator* creates the necessary command files to create the IP block netlists using CORE Generator, invokes CORE Generator, and creates project and script files for HDL simulation, synthesis, technology mapping, placement, routing, and bit-stream generation. To ensure efficient compilation of multi-rate systems, *System Generator* creates constraint files for the physical implementation tools. *System Generator* also creates an HDL test bench for the generated realization, including test vectors computed during Simulink simulation. The end user can use tools such as ModelSim [63] for further verification.

We consider modeling a simple digital filter with the following difference equation in *System Generator*:

$$y_{n+1} = x_{n+1} + a \times x_{n-1} \qquad (2.1)$$

where a is a pre-defined constant parameter. The *System Generator* model for this single-input, single-output system follows directly from this equation and is shown in Figure 2.21.

The filter high-level model is comprised of four *System Generator* blocks: the *Delay* block delays the input data by a specific number of clock cycles; the *Constant* block stores the constant a parameter; the Multiplier block and

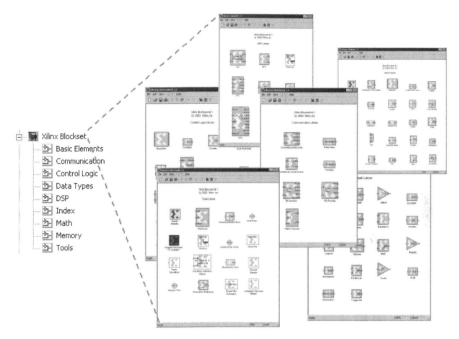

FIGURE 2.20: The high-level Simulink block set provided by *System Generator*

Adder-Subtracter (*AddSub*) block perform the arithmetic calculation on the input data to generate the output data *y*.

2.2.3.2 High-Level Data Type and Rate-Type Propagation

When developing the high-level *System Generator* models, the fix-point data types and the processing rates of the modeling blocks need to be specified. Both signed and unsigned data types (i.e., *Fix* and *UFix* data types) are supported. For example, data type *UFix_24_12* denotes an unsigned fix-point data with length 24 and binary point located at position 12, while *UFix_28_16* denotes a signed fix-point data type with length 28 and binary point located at position 16. When processing data with different types, each of the *System Generator* blocks can automatically convert and align the data types properly before performing the actual computation. If the two inputs to an AddSub block can be *UFix_24_12* and *UFix_28_16*, the AddSub block would align the binary point positions of the input data before performing addition and/or subtraction. Note that the automatic type conversion would be transformed into corresponding low-level implementations, so as to ensure the behavior of the final bitstream matches with the high-level Simulink model.

Each of the Simulink blocks is associated with a processing rate property. This *rate* property is used to denote the operating speed of the *System Gen-*

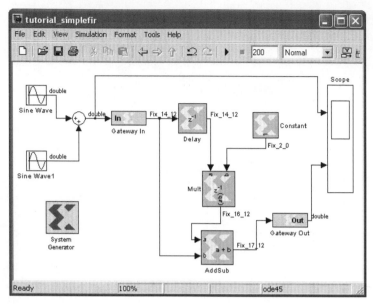

FIGURE 2.21:   An FIR filtering design using *System Generator*

erator block in relation to the input block (in most cases, the simulation clock frequencies of the Simulink model). For example, a block with rate of 2 gets invoked every other Simulink clock. In the corresponding low-level implementation, this clock operates two times slower than the input clock signal. The slower clock is realized using a pair of clock (i.e., *sys_clk*) and clock enable (i.e., *sys_ce*) signals. The concept of processing rates enables the modeling of multi-rate systems, which are widely used in many digital signal processing applications.

To facilitate the modeling of many digital signal processing and image/video processing applications, *System Generator* has the capability to automatically propagate the fix-point data type and the processing rates of the models along the data paths. Considering the examples shown in Figure 2.23, the *empty* output port of the From FIFO block specifies a *Bool* data type, which is propagated along the data path to the following Inverter block and the Register block, etc. Similarly, the processing rate of 1 of the From FIFO block is propagated along the data path to these following blocks.

2.2.3.3   Simulation

Most of the *System Generator* blocks contains cycle-accurate simulation models for the corresponding low-level hardware implementation. For example, the *System Generator* model shown in Figure 2.21 also contains a high-level testbench. The data source is built from Simulink blocks, and is the sum of a high frequency sine wave and a low frequency sine wave. The output of

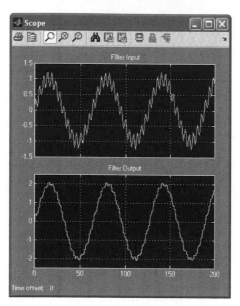

FIGURE 2.22:   Simulation results within the Simulink modeling environment

the simple filter is sent to a Simulink Scope. The System Generator "gateway" blocks convert between Simulink's double-precision floating-point signals and *System Generator*'s fixed-point arithmetic type. After running the simulation in Simulink, the Scope block brings up a graphical display of the input and output of the filter, which is shown in Figure 2.22. It can be readily seen that the filter largely eliminates the high frequency components of the input signal.

In addition to the Simulink simulation models, *System Generator* integrates HDL simulators for co-simulation. Currently, ModelSim from Mentor Graphics [63] and the ISE Simulator from Xilinx [97] are the two officially supported HDL simulators at the time of writing. Through these integrated HDL

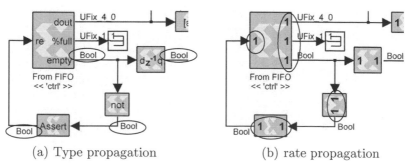

(a) Type propagation   (b) rate propagation

FIGURE 2.23:   Automatic rate and type propagation

simulators, the end users can plug in their low-level designs into the Simulink modeling environment and co-simulate them through these underlying simulators.

Another important co-simulation capability offered by *System Generator* is hardware co-simulation. Using the hardware co-simulation capability, users can choose to implement portions of a *System Generator* and execute it on the actual hardware. *System Generator* automatically generates the communication interface so that the portion of a design running on hardware can communicate with the other portion of the design running on a personal computer. Both single-stepped and free-running simulation modes are provided. In single-stepped mode, the hardware clock is controlled by the personal computer. In each Simulink clock cycle, the design portion running on hardware also advances one clock cycle. The user is presented with a cycle-accurate image of the complete system. Instead, in free-running mode, the design portion mapped to the hardware runs with a "free" clock, which is independent of the design portion simulated on the personal computer. The single-stepped mode reveals much more simulation detail compared with the free-running mode.

Single-stepped mode runs much slower than the free-running mode in most design scenarios. Limited by the computation capability of personal computers, single-stepped mode is too slow for simulating many practical systems. The primary usage of the single-stepped mode is for debugging and/or verifying a specific functional unit of a large system. When mapping a portion of the design properly, free-running mode can greatly increase the overall simulation speed. For example, by mapping the embedded processors onto an actual hardware connected to a personal computer and simulating them through free-running hardware co-simulation, users can effectively simulate the execution of many hardware-software co-design systems.

As more and more complicated systems are developed on reconfigurable hardware platforms, achieving further simulation speed-ups becomes highly desired. For example, video and image processing systems need to process a large amount of data, which require a large amount of computation cycles. Going beyond traditional low-level simulations and single-stepped and free-running hardware, co-simulation modes would be key technologies for successful deployment of these complicated systems.

For both hardware co-simulation and the co-processing interface provided by System Generator, the custom hardware peripherals and the processors communicate with each other through shared memory based memory mapped interface. Herein, shared memory refers to the shared memory blocks in the System Generator high-level hardware design block set, which consists of *From/To Register* blocks, *From/To FIFO* blocks, and the on-chip dual-port memory blocks based *Shared Memory* block. A shared memory block is considered "single" if it has a unique name within a design. When a single shared memory is compiled for hardware co-simulation, System Generator builds the other half of the memory and automatically wires it into the resulting netlist. Additional interfacing logic is attached to the memory that allows it to communicate with the PC. When you co-simulate the shared memory, one half of the memory is used by the logic in your System Generator design. The other half communicates with the PC interfacing logic (Figure 2.24). In this manner, it is possible to communicate with the shared memory embedded inside the FPGA while a simulation is running.

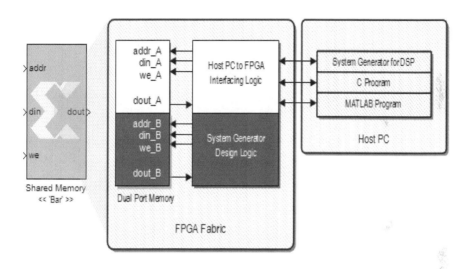

FIGURE 2.24:   Shared memory for hardware co-simulation

The shared memory hardware and interface logic are completely encapsulated by the hardware co-simulation block that is generated for the design. By co-simulating a hardware co-simulation block that contains a shared memory, it is possible for the user design logic and host PC software to share a common address space on the FPGA.

2.2.3.4   Hardware-Software Co-Design

System Generator offers three high-level blocks to allow users to bring various processor subsystems into a high-level design model: the Black Box block, a PicoBlaze Microcontroller block, and an EDK Processor block.

• *Black Box Block*

The Black Box block is provided for importing generic HDL designs into the Simulink modeling environment. The Black Box approach provides the largest degree of flexibility among the three hardware-software co-design approaches at the cost of design complexity. Users can essentially bring in any processor HDL design into a *System Generator* model through the Black Box block. The top-level ports and bus interfaces on the processor can be exposed to the

Black Box block. It is through these exposed ports and bus interfaces that the imported processors interact with other portions of the *System Generator* designs. Since the Black Box block supports the imports of HDL code and netlist files (e.g., *ngc* and *edif* files), users are left with a degree of flexibility to engineer the connectivity between the processor and other *System Generator* blocks. The users have complete control over software compilation issues. However, they need to manually initiate the memory blocks of the imported processors.

Note that the HDL files and netlist files imported through Black Box blocks can be simulated through various HDL simulators. ISE Simulator from Xilinx [97] and ModelSim from Mentor Graphics [63] are the two formally supported HDL simulators in *System Generator*. Since these simulators target hardware development, they may turn out to be very time consuming and even impractical in many software development scenarios. Users can also rely on the hardware co-simulation capability offered by *System Generator* to simulate the processor system imported through Black Box block, which can significantly increase the simulation speed. While generating the low-level implementations and bitstreams of the processor system is a time-consuming process, once the hardware platform is fixed, users can rely on tools such as *data2bram* or *bitinit* to update the software programs running on the processors without having to regenerating the low-level implementations and bitstreams again.

The development of a customized PLB slave hardware peripheral using *System Generator* is presented in [72]. The development flow is shown in Figure 2.25. A basic PowerPC system is described using a simple configuration script, which is then provided to the Xilinx Embedded Development Kit (EDK) to generate the low-level implementations. The related pins on the PLB bus interface are brought up to the top-level ports of the basic processor system. The basic PowerPC system is imported into *System Generator* through a Black Box block. All the top-level ports of the basic PowerPC system are shown up as the input and output ports of the Black Box block. It is through these top-level ports that the *System Generator* model interacts with the basic PowerPC system.

Being able to simulate, and even better, single-step the basic PowerPC system is very useful for the hardware peripheral development in *System Generator*. While there are various instruction-set simulators for simulating the PowerPC processor core and the basic processor system, these simulators suffer from some limitations for practical application development. The PowerPC 405/440 instruction-set simulator provided by IBM [48] does not guarantee full cycle-accuracy. It does not contain simulation models for the Advanced Auxiliary Processor Unit (APU) interface, which is used for tight coupling of hardware peripherals. The PowerPC 405 SmartModel simulation model from Xilinx [97] does not support the initialization of instruction and data cache of the PowerPC processor core. This limitation makes it unable to simulate the PowerPC 405 based ultra-controller applications, which store program instructions and data at the cache. In addition to the limitations of the simulators, a high simulation speed is critical to observe and understand the impact and design trade-offs of different processor system configurations. The low-level behavioral models of the PowerPC systems, such as the pre-compiled PowerPC structural simulation models provided by the Xilinx ISE Simulator, are too slow for practical software development and debugging.

The hardware co-simulation capability provided by *System Generator* supports a variety of popular debugging interfaces (e.g., JTAG) and high-speed communication interfaces (e.g., Ethernet and PCI), and is very easy to set-up. Once a supported hardware prototyping board is connected to the personal computer, users can easily choose the proper compilation target to trigger *System Generator* to automatically generate the hardware bitstreams of the design portions to be run on the hardware board and on the simulation synchronization and data exchange interfaces.

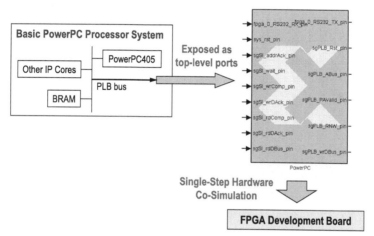

FIGURE 2.25: PowerPC system imported into SysrGen through Black Box block

- *PicoBlaze Micro-controller Block*

The *PicoBlaze Micro-controller* block (referred to as "PicoBlaze block" in the following) provides the smallest degree of flexibility but is the least complex to use among the three hardware-software co-design approaches in *System Generator*. A snapshot of an illustrative example using the PicoBlaze block is shown in Figure 2.26. The *PicoBlaze* block implements an embedded 8-bit microcontroller using the PicoBlaze macro. Both PicoBlaze 2 and PicoBlaze 3 are supported by the PicoBlaze block, depending on the FPGA devices targeted by the end users. The PicoBlaze block exposes a fixed interface to *System Generator*. In the most common design cases, a single block ROM containing 1024 or fewer than 8 bit words serves as the program store. Users can program the PicoBlaze using the PicoBlaze assembler language and let the PicoBlaze block pick up the generated machine code.

The PicoBlaze block integrates a cycle-accurate assembly instruction set simulator for high-level simulation within the Simulink modeling environment. In addition, a *PicoBlaze Instruction Display* block is also provided, which can display the assembly instruction under execution at a specific clock cycle. Users can use the *Single-Step Simulation* block to single-step the PicoBlaze execution and debug the software programs running on PicoBlaze using the *PicoBlaze Instruction Display* block.

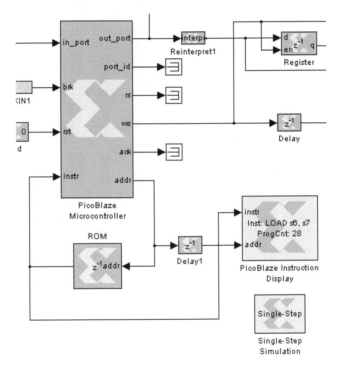

FIGURE 2.26:   Snapshot of a design using the PicoBlaze Block

- *EDK Processor Block*

The EDK Processor block supports two complementary hardware-software co-design flows, which are shown in Figure 2.27. In the Import (denoted as "HDL netlisting" on the block GUI) flow, a MicroBlaze based processor system created using the Xilinx Embedded Development Kit (EDK) is imported as a block of a *System Generator* design. In the Pcore Export flow, a *System Generator* is exported as a customized hardware IP core (also known as *Pcore* in the tool's user manual), which is used by EDK to construct a complete processor system.

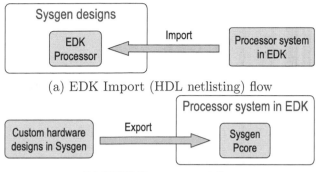

(a) EDK Import (HDL netlisting) flow

(b) EDK Pcore export flow

FIGURE 2.27:   Two complementary flows supported by the EDK
Processor block

For both the Import flow and the Export flow, the EDK Processor block automatically generates a shared memory interface, the architecture of which is illustrated in Figure 2.28. Three types of shared memories are supported: registers, FIFOs (first-in-first-outs), and dual-port block memory blocks. These shared memories are attached to the processor of interest through a bus adapter, which maps the shared memories to different locations of the processor memory space. The bus adapter currently supports both the dedicated FSL interface and the shared PLB interface. For the FSL interface, the bus adapter can be attached to any EDK system with FSL interfaces, such as MicroBlaze and PowerPC with APU-FSL bridges. For the PLB interface, the bus adapter can be attached to PLB-enabled MicroBlaze and PowerPC processors. In addition to the low-level hardware implementations, software drivers for accessing the shared memory interface are also generated automatically. Figure 2.29 depicts the adding of a shared memory block onto the MicroBlaze memory map interface.

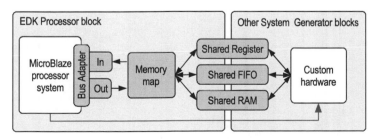

FIGURE 2.28:   Automatically generated memory map interface

In order to provide more design flexibility, the EDK Processor block also

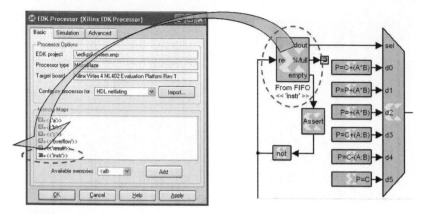

FIGURE 2.29:   Adding shared memories to MicroBlaze memory map

supports the direct exposing of top-level ports of the imported EDK process systems. Users can use the exposed port feature to implement customized bus interfaces that are not supported by the EDK Processor block bus adapter to communicate with the imported processor. Figure 2.30 shows a design case where the OPB bus interface is exposed as the input and output ports of the EDK Processor block. Users can implement OPB decoding themselves in order to talk to the imported MicroBlaze processor. See [10] for more details about developing customized OPB (on-chip peripheral bus) interface using *System Generator*.

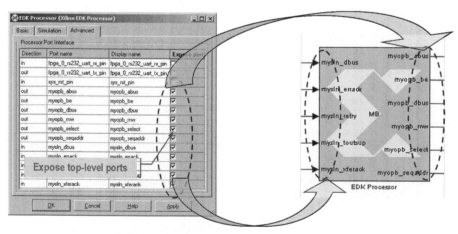

FIGURE 2.30:   Exposed top-level ports

System Generator provides a high-level event generation interface to allow users to specify the events of the shared memory interface that they are interested in. A screen-shot of the graphical interface is shown in Figure 2.31. Through the graphical interface, users can create an event that is triggered when a specific value appears on a shared register. The processor is notified of the triggered event as an incoming external interrupt. Users can write an interrupt service routine (ISR) to properly process the triggered event.

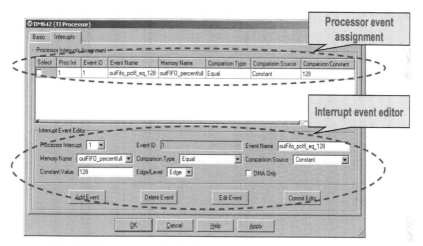

FIGURE 2.31: High-level event generation interface

To facilitate software development on the imported processors, users can perform co-debugging using software debuggers (e.g., the GNU GDB debugger). The screen-shot in Figure 2.32 shows that the Simulink simulation is paused when the execution on the imported MicroBlaze reaches a breakpoint. Users can then examine the status of the high-level model through Simulink scopes or the *WaveScope* provided by *System Generator*, or examine the status of the processor through GDB. In the case of software based simulation, the imported MicroBlaze is simulated using a cycle-accurate instruction set simulator. There are virtual I/O console for displaying the processor output to general-purpose I/O ports and UART serial consoles.

2.2.3.5 Performance Analysis and Resource Estimation

Identifying critical paths and getting timing closure are important for practical application development. Since the *System Generator* blocks many abstract way low-level implementation details, users may have difficulties when they need to fix designs with timing violations. To address this issue, a *Timing Analyzer* tool is provided to help achieve timing goals.

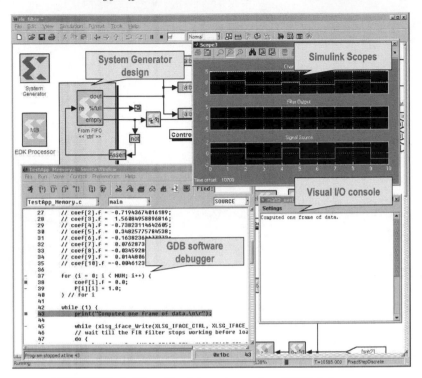

FIGURE 2.32: Co-debugging using GDB debugger

Timing Analyzer first generates the corresponding low-level implementations of the high-level design model. Then it invokes *Trace*, a low-level timing analysis tool provided by the Xilinx ISE tool [97], to analyze the timing information and identify critical paths of the low-level implementations. The report from *Trace* is then parsed to achieve timing information for the high-level modeling blocks. There are naming correlations between the high-level models and the low-level implementations. *Timing Analyzer* relies on this naming correlations to find the high-level blocks that are on the critical paths of the low-level implementations, and it then highlights these high-level blocks. Based on this information, users can perform optimizations, such as extra pipelining and register re-timing, to meet the design timing budget.

Using a similar approach as the *Timing Analyzer* tool, *System Generator* also provides a *Resource Estimator* tool. *Resource Estimator* can report the hardware resource usage of the high-level blocks by analyzing low-level synthesis reports and/or place-and-route information. The naming correlation between the high-level model and the corresponding low-level implementations is used to generate resource usage estimation for the high-level modeling blocks.

Note that the results given by the *Timing Analyzer* tool and the *Resource*

Estimator tool are estimates only, not exact values. This is due to the fact that the process for generating the low-level implementations is a very complicated process. On the one perspective, the synthesis tools and the place-and-route tools may trim off or combine some logic components and wires for optimization purposed. The low-level implementations of two high-level blocks may be combined, making it impossible to have accurate resource utilization values for these blocks. Register re-timing may move a register to other locations of the data path, which may not be correlated with the high-level Register block. On the other hand, the naming correlation of the high-level blocks may be lost during the optimization process. Thus, there are some low-level components that the corresponding high-level blocks cannot easily identify by their names.

Chapter 3

A High-Level Hardware-Software Application Development Framework

3.1 Introduction

Paradoxically, while FPGA based configurable multi-processor platforms offer a high degree of flexibility for application development, performing design space exploration on configurable multi-processor platforms is very challenging. State-of-the-art design tools rely on *low-level simulation*, which is based on the register transfer level (RTL) and/or gate level implementations of the platform, for design space exploration. These low-level simulation techniques are inefficient for exploring the various hardware and software design trade-offs offered by configurable multi-processor platforms. This is because of two major reasons. One reason is that low-level simulation based on register transfer/gate level implementations is too time consuming for evaluating the various possible configurations of the FPGA based multi-processor platform and hardware-software partitioning and implementation possibilities. Especially, RTL and gate-level simulation is inefficient for simulating the execution of software programs running on configurable multi-processor platforms. Considering the design examples shown in Section 3.5.1.2, low-level simulation using ModelSim [63] takes more than 25 minutes to simulate an FFT computation software program running on the MicroBlaze processor with a 1.5 msec execution time. This simulation speed can be overwhelming for development on configurable multi-processor platforms as software programs usually take minutes or hours to complete on soft processors. The other reason is that FPGA based configurable multi-processor platforms pose a potentially vast design space and optimization possibilities for application development. There are various hardware-software partitions of the target application and various possible mappings of the application to the multiple processors. For customized hardware development, there are many possible realizations of the customized hardware peripherals depending on the algorithms, architectures, hardware bindings, etc., employed by these peripherals. The communication interfaces between various hardware components (e.g., the topology for connecting the processors, the communication protocols for exchanging data between the processors and the customized hardware peripherals and among the processors) would also significantly make efficient design space exploration more challenging. Exploring such a large design space using time-consuming low-level simulation techniques becomes intractable.

We address the following design problem in this section. The target application is composed of a number of tasks with data exchange and execution precedence between each other. Each of the tasks is mapped to one or more processors for execution. The hardware architecture of the configurable multi-processor platform of interest is shown in Figure 3.1. Application development on the multi-processor platform involves both software designs and hardware designs. For software designs, the application designer can choose the mapping and scheduling of the tasks for distributing the software programs of the tasks among the processors so as to process the input data in parallel. For hardware designs, there are two levels of customization that can be done to the hardware platform. On the one hand, customized hardware peripherals can be attached to the soft processors as hardware accelerators to speed up some computation steps. Different bus interfaces can be used for exchanging data between the processors and their own hardware accelerators. On the other hand, the processors are connected through a communication network for cooperative data processing. Various topologies and communication protocols can be used when constructing the communication network. Based on the above assumptions, our objective is to *build an environment for design space exploration on a configurable multi-processor platform that has the following desired properties. (1) The various hardware and software design possibilities offered by the multi-processor platform can be described within the design environment. (2) For a specific realization of the application on the multi-processor platform, the hardware execution (i.e., the execution within the customized hardware peripheral and the communication interfaces) and the software execution within the soft processors are simulated rapidly in a concurrent manner so as to facilitate the exploration of the various hardware-software design flexibilities. (3) The results gathered during the hardware-software co-simulation process should facilitate the identification of the candidate designs and diagnose of performance bottlenecks. Finally, after the candidate designs are found out through the co-simulation process, the corresponding low-level implementations with the desired high-level behavior can be generated automatically.*

As the main contribution of this chapter, we demonstrate that *a design space exploration approach based on arithmetic level cycle-accurate hardware-software modeling and co-simulation can achieve the objective described above.* As discussed in Section 6.5, on the one hand, the arithmetic level modeling and co-simulation technique does not involve the low-level RTL and/or gate level implementations of the multi-processor platform and thus can greatly increase the simulation speed. It enables efficient exploration of the various design trade-offs offered by the configurable multi-processor platform. On the other hand, maintaining cycle-accuracy during the co-simulation process helps the application designer to identify performance bottlenecks and thus facilitate the

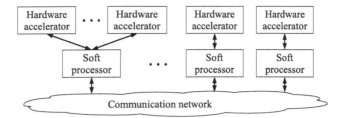

FIGURE 3.1: Hardware architecture of the configurable multi-processor platform

identification of candidate designs with "good" performance. It also facilitates the automatic generation of low-level implementations once the candidate designs are identified through the arithmetic level co-simulation process. We consider performance metrics of time and hardware resource usage in this chapter. For illustrative purposes, we show an implementation of the proposed arithmetic level co-simulation technique based on MATLAB/Simulink [60]. Through the design of three widely used numerical computation and image processing applications, we show that the proposed cycle-accurate arithmetic level co-simulation technique achieves speed-ups in simulation time up to more than 800x as compared to those achieved by low-level simulation techniques. For these three applications, the designs identified using our co-simulation environment achieve execution speed-ups up to more than 5.6x compared with other designs considered in our experiments. Finally, we have verified the low-level implementations generated from the arithmetic level design description on a commercial FPGA prototyping board.

This chapter is organized as follows. Section 3.2 discusses related work. Section 3.3 describes our approach for building an arithmetic level cycle-accurate co-simulation environment. An implementation of the co-simulation environment based on MATLAB/Simulink is provided in Section 3.4 to illustrate the proposed co-simulation technique. Two signal processing applications and one image processing application are provided in Section 3.5 to demonstrate the effectiveness of our co-simulation technique. Finally, we conclude in Section 3.6.

3.2   Related Work

Many previous techniques have been proposed for performing hardware-software co-design and co-simulation on FPGA based hardware platforms. Hardware-software co-design and co-simulation on FPGAs using compiler optimization techniques are proposed by Gupta *et al.* [37], Hall *et al.* [39], Palem *et al.* [73], and Plaks [74]. A hardware-software co-design technique for a data-driven accelerator is proposed by Hartensten and Becker [41]. To our best knowledge, none of the prior work addresses the rapid design space exploration problem for configurable multi-processor platforms.

State-of-the-art hardware/software co-simulation techniques can be roughly classified into four major categories, which are discussed below.

• *Techniques based on low-level simulation*: Since configurable multi-processor platforms are configured using FPGA resources, the hardware-software co-simulation of these platforms can use low-level hardware simulators directly. This technique is used by several commercial integrated design environments (IDEs) for application development using configurable multi-processor platforms. This includes SOPC Builder from Altera [3] and Embedded Development Kit (EDK) from Xilinix [97]. When using these IDEs, the multi-processor platform is described using a simple configuration script. These IDEs will then automatically generate the low-level implementations of the multi-processor platform and the low-level simulation models based on the low-level implementations. Software programs are also compiled into binary executable files, which are then used to initialize the memory blocks in the low-level simulation models. Based on the low-level simulation models, low-level hardware simulators (e.g., [63]) can be used to simulate the behavior of the complete multi-processor platform. From one standpoint, the simple configuration scripts used in these IDEs provide very limited capabilities for describing the two types of optimization possibilities offered by configurable multi-processor platforms. From another standpoint, as we show in Section 3.5, such a low-level simulation based co-simulation approach is too time-consuming for simulating the various possibilities of application development on the multi-processor platforms.

• *Techniques based on high-level languages*: One approach of performing hardware-software co-simulation is by adopting high-level languages such C/C++ and Java. When applying these techniques, the hardware-software co-simulation can be performed by compiling the designs using their high-level language compilers and running the executable files resulting from the compilation process. There are several commercial tools based on C/C++. Examples of such co-simulation techniques include Catapult C from [62] and Impulse C, which is used by the CoDeveloper design tool from [47]. In addition to supporting the standard ANSI/ISO C, both Catapult C and Impulse C provide language extensions for specifying hardware implementation properties. The application designer describes his/her designs using these extended C/C++ languages, compiles the designs using standard C/C++ compilers, generates the binary executable files, and verifies the functional behavior of the designs by analyzing the output of the executable files. To obtain the cycle-accurate functional behavior of the designs, the application designer still needs to generate the VHDL simulation models of the designs and perform low-level simulation using cycle-accurate hardware simulators. The DK3 tool from [15] supports development on configurable platforms using Handel-C [16] and SystemC [67], both extensions of C/C++ language. Handel-C and SystemC allow for the description of hardware and software designs at different abstraction levels. However, to make a design described using Handel-C or SystemC suitable for direct register transfer level generation, the application designer needs to write designs at nearly the same level of abstraction as handcrafted register transfer level hardware implementations [61]. This would prevent efficient design space exploration for configurable multi-processor platforms.

• *Techniques based on software synthesis*: For software synthesis based hardware-software co-simulation techniques, the input software programs are synthesized into Co-design Finite State Machine (CFSM) models which demonstrate the same functional behavior as that of the software programs. These CFSM models are then integrated into the simulation models for the hardware platform for hardware-software co-simulation. One example of the software synthesis approach is the POLIS hardware-software co-design framework [9]. In POLIS, the input software programs are synthesized and translated into VHDL simulation models which demonstrate the same functional behavior. The hardware simulation models are then integrated with the simulation models of other hardware components for co-simulation of the complete hardware platform. The software synthesis approach can greatly accelerate the time for co-simulating the low-level register transfer/gate level implementations of multi-processor platforms. One issue with the software synthesis co-simulation approach is that it is difficult to synthesize complicated software programs (e.g., operating systems, video encoding/decoding software programs) into hardware simulation models with the same functional behavior. Another issue is that the software synthesis approach is based on low-level implementations of the multi-processor platform. The application designer needs to generate the low-level implementations of the complete multi-processor platform before he/she can perform co-simulation. The large amount of effort required by generating the low-level implementations prevents efficient exploration of various configurations of the configurable multi-processor platform.

• *Techniques based on integration of low-level simulators*: Another approach for hardware-software co-simulation is through the integration of low-level hardware-software simulators. One example is the *Seamless* tool from [63]. It contains pre-compiled cycle-accurate behavioral processor models for simulating the execution of software programs on various processors. It also provides pre-compiled cycle-accurate bus models for simulating the communication between hardware and software executions. Using these pre-compiled processor models and bus models, the application designer can separate the simulation of hardware executions and software executions and use the corresponding hardware and software simulators for co-simulation. The *Seamless* tool provides detailed simulation information for verifying the correctness of hardware and software co-designs. However, it is still based on the time-consuming low-level simulation techniques. Therefore, this low-level simulation based technique is inefficient for application development on configurable multi-processor plat-

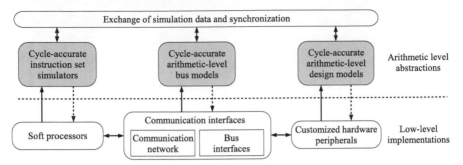

FIGURE 3.2: Our approach for high-level hardware-software
co-simulation

forms.

To summarize, the design approaches discussed above rely on low-level simulation to obtain the cycle-accurate functional behavior of the hardware-software execution on configurable multi-processor platforms. Such reliance on low-level simulation models prevents them from efficiently exploring the various configurations of multi-processor platforms. Compared with these approaches, the arithmetic level co-simulation technique proposed in this chapter allows for arithmetic level cycle-accurate hardware-software co-simulation of the multi-processor platform without involving the low-level implementations and low-level simulation models of the complete platform. Our approach is able to achieve a significant simulation speed-up while providing cycle-accurate simulation details compared with low-level simulation techniques.

3.3   Our Approach

Our approach for design space exploration is based on an arithmetic level co-simulation technique. In the following paragraphs, we present the arithmetic level co-simulation technique. Design space exploration for multi-processor platform can be performed by analyzing the arithmetic level co-simulation results.

• *Arithmetic level modeling and co-simulation*: The arithmetic level modeling and co-simulation technique for configurable multi-processor platforms are shown in Figure 3.2. The configurable multi-processor platform consists of three major components: *soft processors* for executing software programs; *customized hardware peripherals* as hardware accelerators for parallel execution of some specific computation steps; and *communication interfaces* for data exchange between various hardware components. The communication interfaces include *bus interfaces* for exchanging data between the processors and their customized hardware peripherals, and *communication networks* for coordinating the computations and communication among the soft processors. When employing our arithmetic level co-simulation technique, arithmetic level ("high-level") abstractions are created to model each of the three major components. The arithmetic level abstractions can greatly speed up the co-simulation process while allowing the application designer to explore the optimization opportunities provided by configurable multi-processor platforms. By "arithmetic level", we mean that only the arithmetic aspects of the hardware and software execution are modeled by these arithmetic abstractions. Taking multiplication as an example, its low-level implementation on Xilinx Virtex-II/Virtex-II Pro FPGAs can be realized using either slice-based multipliers or embedded multipliers. Its arithmetic level abstraction only capture the arithmetic property, i.e., multiplication of the values presented at its input ports. Taking the communication between different hardware components as another example, its low-level implementations can use registers (flip-flops), slices, or embedded memory blocks to realize data buffering. Its arithmetic level abstraction only capture the arithmetic level data movements on the communication channels (e.g., the handshaking protocols, access priorities for communication through shared communication channels). Co-simulation based on the arithmetic level abstractions of the hardware components does not involve the low-level implementation details. Using the arithmetic-level abstraction, the user can specify that the movement of the data be delayed for one clock cycle before it becomes available to the destination hardware components. He/she does not need to provide low-level implementation details (such as the number of flip-flop registers and the connections of the registers between the related hardware components) for realizing this arithmetic operation. Such low-level implementations can be generated automatically once the high-level design description is finished. Thus, these arithmetic-level abstractions can significantly speed up the time required to simulate the hardware and software arithmetic behavior of the multi-processor platform. The implementation of the arithmetic level co-simulation technique presented in Section 3.4 demonstrates a simulation speed-up up to more than 800x compared with the behavioral simulation based on low-level implementations of the hardware platforms.

The configurable multi-processor platform is described using the arithmetic level abstractions. The arithmetic level abstractions allow the application designer to specify the various ways of constructing the data paths through which the input data is processed. Co-simulation based on the arithmetic level abstractions gives the status of the data paths during the execution of the application. As an example, the development of a JPEG2000 application on a state-of-the-art configuration multi-processor platform using the arithmetic level abstractions is shown in Section 3.5.2. The arithmetic level abstraction of the JPEG2000 application specifies the data paths that the input image data are processed among the multiple processors. The co-simulation based on the arithmetic level abstraction gives the intermediate processing results at various hardware components that constitute the multi-processor platform. Using these intermediate processing results, the application designer can analyze performance bottlenecks and identify the candidate designs of the multi-processor platform. Application development using the arithmetic level abstractions focus on the description of data path. Thus, the proposed arithmetic level co-simulation technique is especially suitable for the development of data-intensive applications such as signal and image processing applications.

The arithmetic level abstractions of the configurable multi-processor platform are simulated using their corresponding hardware and software simulators. These hardware and software simulators are tightly integrated into our co-simulation environment and concurrently simulate the arithmetic behavior of the complete multi-processor platform. Most importantly, the simulations performed within the integrated simulators are synchronized between each other at each clock cycle and provide cycle accurate simulation results for the complete multi-processor platform. By "cycle-accurate", we mean that for each clock cycle during the co-simulation process, the arithmetic behavior of the multi-processor platform simulated by the proposed co-simulation environment matches with the arithmetic behavior of the corresponding low-level implementations. For example, when simulating the execution of software programs, the cycle-accurate co-simulation takes into account the number of clock cycles required by the soft processors for completing a specific instruction (e.g., the multiplication instruction of the MicroBlaze processor takes three clock cycles to complete). When simulating the hardware execution on customized hardware peripherals, the cycle-accurate co-simulation takes into account the number of clock cycles required by the pipelined customized hardware peripherals to process the input data. The cycle-accurate co-simulation process also takes into account the delays caused by the communication channels between the various hardware components. One approach for achieving such cycle accuracy between different simulators in the actual implementations is by maintaining a global simulation clock in the co-simulation environment. This global simulation clock can be used to specify the time at which a piece of data is available for processing by the next hardware or software component based on the delays specified by the user. Maintaining such cycle-accurate property in the arithmetic level co-simulation ensures that the results from the arithmetic level co-simulation are consistent with the arithmetic behavior of the corresponding low-level implementations. The cycle-accurate simulation results allow the application designer to observe the instant interactions between hardware and software executions. The instant interaction information is used in the design space exploration process for identifying performance bottlenecks in the designs.

• *Design space exploration*: Design space exploration is performed based on the arithmetic level abstractions. The functional behavior of the candidate designs can be verified using the proposed arithmetic level co-simulation technique. As shown in Section 3.5, the application designer can use the simulation results gathered during the cycle-accurate arithmetic level co-simulation process to identify the candidate designs and diagnose the performance bottlenecks when performing design space exploration. Considering the development of a JPEG2000 application on a multi-processor platform discussed in Section 3.5.2, using the results from the arithmetic level simulation can identify that the bus connecting the multiple processors limits the performance of the complete system when more processors are employed. Besides, these detailed simulation results also facilitate the automatic generation of the low-level implementations. By specifying the low-level hardware bindings for the arithmetic operations (e.g., binding the embedded multipliers for realization of the multiplication arithmetic operation), the application designer can also rapidly obtain the hardware resource usage for a specific realization of the application [84]. The cycle-accurate arithmetic level simulation results and the rapidly estimated hardware resource usage information can help the application designer efficiently explore the various optimization opportunities and identify "good" candidate designs. For example, in the development of a block matrix multiplication algorithm shown in Section 3.5.1.1, the application designer can explore the impact of the size of the matrix blocks on the performance of the complete algorithm. Finally, for the designs identified by the arithmetic level co-simulation process, low-level implementations with corresponding arithmetic behavior are automatically generated based on the arithmetic level abstractions.

3.4   An Implementation Based on MATLAB/Simulink

For illustrative purposes, we provide an implementation of our arithmetic level co-simulation approach based on MATLAB/Simulink for application development using configurable multi-processor platforms. The software architecture of the implementation is shown in Figure 3.3 and Figure 3.4. Arithmetic level abstractions of the customized hardware peripherals and the communication interfaces (including the bus interfaces and the communication network) are created within the MATLAB/Simulink modeling environment. Thus, the hardware execution platform is described and simulated within MATLAB/Simulink. Additionally, we create *soft processor Simulink blocks* for integrating cycle-accurate instruction set simulators targeting the soft processors. The execution of the software programs distributed among the soft processors is simulated using these soft processor Simulink blocks.

3.4.1   High-Level Design Description

Our hardware-software development framework provides a new block called *Software Co-Simulation* block, in the Simulink modeling environment. This new *Software Co-Simulation* block unifies the application development capabilities offered by both the System Generator tool and the EDK (Embedded Development Kit) tool into the MATLAB/Simulink modeling environment. More specifically, by offering the *Software Co-Simulation* block, the proposed framework presents the end user with the following two high-level abstractions for rapid construction of the hardware-software execution systems.

1. Arithmetic-level abstractions for hardware platform development: The end user can describe the arithmetic-level behavior of the hardware platform using the Simulink blocks provided by the *System Generator* tool. For example, during the high-level arithmetic-level modeling, the user can specify that the presence of a value to its destination Simulink blocks be delayed for one clock cycle. The corresponding low-level implementation is composed of a specific number of flip-flop registers based on the data type of the high-level signal and the wires that connect them properly to the low-level implementations of other components. The application framework automatically generates the corresponding low-level implementation.

2. Interfacing level abstractions for configuration of the processor systems: EDK provides a simple script based MHS (Microprocessor Hardware System) file for describing the connections between the hardware peripherals of the processors. The end user can also specify the setting of the processor (e.g. cache size and memory mapping, etc.) and the hardware peripherals of the processor within the MHS file.

3.4.2   Arithmetic-Level Co-Simulation

Our arithmetic level co-simulation environment consists of four major components: *simulation of software execution on soft processors, simulation of customized hardware peripherals, simulation of communication interfaces*, and *exchange of simulation data and synchronization*. They are discussed in detail in the following subsections.

3.4.2.1   Simulation of Software Execution on Soft Processors

Soft processor Simulink blocks (e.g., MicroBlaze Simulink blocks targeting MicroBlaze processors) are created for simulating the software programs running on the processors. Each soft processor Simulink block simulates the software programs executed on one processor. Multiple soft processor Simulink blocks are employed in order to properly simulate the configurable multiprocessor platform.

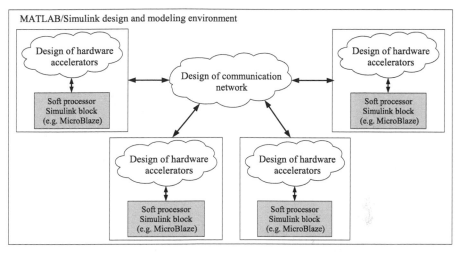

FIGURE 3.3: An implementation of the proposed arithmetic co-simulation environment based on MATLAB/Simulink

The software architecture of a soft processor Simulink block is shown in Figure 3.4. The input C programs are compiled using the compiler for the specific processor (e.g., the GNU C compiler *mb-gcc* for MicroBlaze) and translated into binary executable files (e.g., *.ELF* files for MicroBlaze). These binary executable files are then simulated using a cycle-accurate instruction set simulator for the specific processor. Taking the MicroBlaze processor as an example, the executable *.ELF* files are loaded into *mb-gdb*, the GNU C debugger for MicroBlaze. A cycle-accurate instruction set simulator for the MicroBlaze processor is provided by Xilinx. *mb-gdb* sends instructions of the loaded executable files to the MicroBlaze instruction set simulator and performs cycle-accurate simulation of the execution of the software programs. *mb-gdb* also sends/receives commands and data to/from MATLAB/Simulink through the soft processor Simulink block and interactively simulates the execution of the software programs in concurrence with the simulation of the hardware designs within MATLAB/Simulink.

3.4.2.2 Simulation of Customized Hardware Peripherals

The customized hardware peripherals are described using the MATLAB/Simulink based FPGA design tools. For example, *System Generator* supplies a set of dedicated Simulink blocks for describing parallel hardware designs using FPGAs. These Simulink blocks provide arithmetic level abstractions of the low-level hardware components. There are blocks that represent the basic hardware resources (e.g., flip-flop based registers, multiplexers), blocks that represent control logic, mathematical functions, and memory, and blocks that represent proprietary IP (Intellectual Property) cores (e.g., the IP cores for

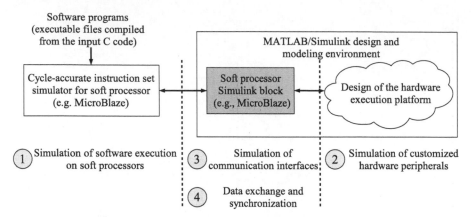

FIGURE 3.4: Architecture of the soft processor Simulink block

Fast Fourier Transform and finite impulse filters). The *Mult* Simulink block for multiplication provided by *System Generator* captures the arithmetic behavior of multiplication by presenting at its output port the product of the values presented at its two input ports. Whether the low-level implementation of the *Mult* Simulink block is realized using embedded or slice-based multipliers is ignored in its arithmetic level abstraction. The application designer assembles the customized hardware peripherals by dragging and dropping the blocks from the block set to the designs and connecting them via the Simulink graphic interface. Simulation of the customized hardware peripherals is performed within MATLAB/Simulink. MATLAB/Simulink maintains a simulation timer for keeping track of the simulation process. Each unit of simulation time counted by the simulation timer corresponds to one clock cycle experienced by the corresponding low-level implementations.

3.4.2.3 Simulation of Communication Interfaces

As shown in Figure 3.3, the simulation of the communication interfaces is composed of two parts, simulation of the bus interfaces between the processors and their corresponding peripherals and simulation of the communication network between the processors, which are described in the following paragraphs.

• *Simulation of the dedicated bus interfaces*: Simulation of the dedicated bus interfaces between the processors and their hardware peripherals is performed by the soft processor Simulink block. Basically, the soft processor Simulink blocks need to simulate the input/output communication protocols and the data buffering operations of the dedicated bus interfaces.

We use MicroBlaze processors and the dedicated Fast Simplex Link (FSL) bus interfaces to illustrate the co-simulation process. FSLs are uni-directional FIFO (First-In-First-Out) channels. As shown in Figure 3.5, each MicroBlaze processor provides up to 16 pairs of dedicated FSL bus interfaces (one for

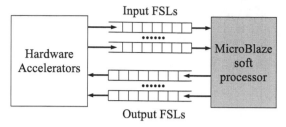

FIGURE 3.5: Communication between MicroBlaze and customized hardware designs through Fast Simplex Links

data input and one for data output in each FSL pair) for direct access of its register file. Customized hardware can be tightly coupled with the MicroBlaze processor through these FSL interfaces. It takes two clock cycles for data to move in and out of the MicroBlaze processor register files through these FSL bus interfaces. Using these FSL channels, the application designer can attach customized hardware peripherals as hardware accelerators to the MicroBlaze processor. Both synchronous (blocking) and asynchronous (non-blocking) read/write operations are supported by MicroBlaze. For blocking read/write operations, the MicroBlaze processor is stalled until the read/write operations are completed. For non-blocking read/write operations, MicroBlaze resumes its normal execution immediately after the read/write operations, regardless of whether these operations are successful or not.

The MicroBlaze Simulink blocks simulate the FSL FIFO buffers and the bus interfaces between the customized hardware peripherals and the FSL channels. The width of the FSL channels is 32 bit while their depths are configurable from 1 to 8192 depending on the application requirements and the available hardware resources on the FPGA device. Let "#" denote the ID of the FSL input/output channel for accessing the MicroBlaze processor. "#" is an integer number between 0 and 7. When the *In#_write* input port of the MicroBlaze Simulink block becomes high, it indicates that there is data from the customized hardware peripherals simulated in MATLAB/Simulink. The data will be written into the FSL FIFO buffer stored at the internal data structure of the MicroBlaze Simulink block. The MicroBlaze Simulink block would then store the MATLAB/Simulink data presented at the *In#_write* input port into the internal data structure and raises the *Out#_exists* output port stored in its internal data structure to indicate the availability of the data. Similarly, when the FSL FIFO buffer is full, the MicroBlaze Simulink block will raise the *In#_full* output port in MATLAB/Simulink to prevent further data from coming into the FSL FIFO buffer.

The MicroBlaze Simulink blocks also simulate the bus interfaces between the MicroBlaze processor and the FSL channels. A set of dedicated C functions is provided for the MicroBlaze to control the communication through FSLs. After compilation using *mb-gcc*, these C functions are translated to the corresponding dedicated assembly instructions. For example, C function *microblaze_nbread_datafsl(val,id)* is used for non-blocking reading of data from the *id*-th FSL channel. This C function is translated into a dedicated assembly instruction *nget(*val*, rfsl#id)* during the compilation of the software program. By observing the execution of these dedicated C functions during co-simulation, we can control the hardware and software processes to correctly simulate the interactions between the processors and their hardware peripherals.

During the simulation of the software programs, the MicroBlaze Simulink block keeps track of the status of the MicroBlaze processor by communicating with *mb-gdb*. As soon as the dedicated assembly instruction described above for writing data to the customized hardware peripherals through the FSL channels is encountered by the *mb-gdb*, it informs the MicroBlaze Simulink block. The MicroBlaze Simulink block will then stall the simulation of software programs in *mb-gdb*, extract the data from *mb-gdb*, and try to write the data into the FSL FIFO buffer stored at its internal data structure. For communication in the blocking mode, the MicroBlaze Simulink block stalls simulation of the software programs in *mb-gdb* until the write operation is completed. That is, the simulation of software programs gets stalled until the *In#_full* flag bit stored at the MicroBlaze Simulink block internal data structure becomes low. This indicates that the FSL FIFO buffer is ready to accept more data. Otherwise, for communication in non-blocking mode, the MicroBlaze Simulink block resumes the simulation of the software programs immediately after writing data to the FSL FIFO buffer regardless of the outcome of the write operation. Data exchange for the read operation is handled in a similar manner.

• *Simulation of the communication network*: Each soft processor and its customized hardware peripherals are developed as a *soft processor subsystem* in MATLAB/Simulink. In order to provide the flexibility for exploring the different topologies and communication protocols for connecting the processors, the communication network that connects the *soft processor subsystems* and coordinates the computations and communication between these subsystems is described and simulated within MATLAB/Simulink. As shown in the design example discussed in Section 3.5.2, an OPB (On-chip Peripheral Bus) Simulink block is created to realize the OPB shared bus interface. The multiple MicroBlaze *subsystems* are connected to the OPB Simulink block to form a bus topology. A hardware semaphore (i.e., a mutual exclusive access controller) is used to coordinate the hardware-software execution among the multiple MicroBlaze processors. The hardware semaphore is described and simulated within MATLAB/Simulink.

3.4.2.4 Exchange of Simulation Data and Synchronization between the Simulators

The soft processor Simulink blocks are responsible for exchanging simulation data between the software and hardware simulators. The input and output ports of the soft processor Simulink blocks are used to separate the simulation of the software programs running on the soft processor and that of the other Simulink blocks (e.g., the hardware peripherals of the processor as well as other soft processors employed in the design). The input and output ports of the soft processor Simulink blocks correspond to the input and output ports of the low-level hardware implementations. For low-level ports that are both input and output ports, they are represented as separate input and output blocks suffixed with port names "_in" and "_out" respectively on the Simulink blocks. The MicroBlaze Simulink blocks send the values of the FSL registers at the MicroBlaze instruction set simulator to the input ports of the soft processor Simulink blocks as input data for the hardware peripherals. Vice versa, the MicroBlaze Simulink blocks collect the simulation output of the hardware peripherals from the output ports of the soft processor Simulink blocks and use the output data to update the values of the FSL registers stored at its internal data structure.

When exchanging the simulation data between the simulators, the soft processor Simulink blocks take into account the number of clock cycles required by the processors and the customized hardware peripherals to process the input data. They also take into account the delays caused by transmitting the data through the dedicated bus interfaces and the communication network. By doing so, the hardware and software simulation are synchronized on a cycle accurate basis.

Moreover, a global simulation timer is used to keep track of the simulation time of the complete multi-processor platform. All hardware and software simulations are synchronized with this global simulation timer. For the MATLAB/Simulink based implementation of the co-simulation environment, one unit of simulation time counted by the global simulation timer equals one unit of simulation time within MATLAB/Simulink and one clock cycle simulation time of the MicroBlaze instruction set simulator. It is ensured that one unit of the simulation time counted by the global simulation timer also corresponds to one clock cycle experienced by the corresponding low-level implementations of the multi-processor platform.

3.4.3   Rapid Hardware Resource Estimation

Being able to rapidly obtain the hardware resources occupied by various configurations of the multi-processor platform is required for design space exploration. For Xilinx FPGAs, we focus on the number of slices, the number of BRAM memory blocks, and the embedded 18-bit-by-18-bit multipliers used for constructing the multi-processor platform.

For the multi-processor platform based on MicroBlaze processors, the hardware resources are used by the following four types of hardware components: the MicroBlaze processors, the customized hardware peripherals, the communication interfaces (the dedicated bus interfaces and the communication network), and the storage of the software programs. Resource usage of the MicroBlaze processors, the two LMB (Local Memory Bus) interface controllers, and the dedicated bus interfaces is estimated from the Xilinx data sheet. Resource usage of the customized hardware designs and the communication network is estimated using the resource estimation technique provided by Shi et al. [84]. Since the software programs are stored in BRAMs, we obtain the size of the software programs using the *mb-objdump* tool and then calculate the numbers of BRAMs required to store these software programs. The resource usage of the multi-processor platform is obtained by summing up the hardware resources used by the four types of hardware components mentioned above.

3.5   Illustrative Examples

To demonstrate the effectiveness of our approach, we show in this section the development of two widely used numerical computation applications (i.e., CORDIC algorithm for division and block matrix multiplication) and one image processing application (i.e., JPEG2000 encoding) on a popular configurable multi-processor platform. The two numerical computation applications are widely deployed in systems such as radar systems and software defined radio systems [57]. Implementing these applications using soft processors provides the capability of handling different problem sizes depending on the specific application requirements.

We first demonstrate the co-simulation process of an individual soft processor and its customized hardware peripherals. We then show the co-simulation process of the complete multi-processor platform, which is constructed by connecting the individual soft processors with customized hardware peripherals through a communication network. Our illustrative examples focus on the MicroBlaze processors and the FPGA design tools from Xilinx due to their wide availability. Our co-simulation approach is also applicable to other soft processors and FPGA design tools. Virtex-II Pro FPGAs [97] are chosen as our target devices. Arithmetic level abstractions of the hardware execution platform are provided by *System Generator* 8.1EA2. Automatic generation of the low-level implementations of the multi-processor platform is realized using both *System Generator* 8.1EA2 and EDK (Embedded Development Kit) 7.1. The ISE (Integrated Software Environment) 7.1 [97] is used for synthesizing and implementing (including placing-and-routing) the complete multi-processor platform. Finally, the functional correctness of the multi-processor platform is verified using an ML300 Virtex-II Pro prototyping board from Xilinx [97].

3.5.1 Co-Simulation of the Processor and Hardware Peripherals

3.5.1.1 Adaptive CORDIC Algorithm for Division

The CORDIC (COordinate Rotation DIgital Computer) iterative algorithm for dividing b by a [6] is described as follows. Initially, we set $X_{-1} = a$, $Y_{-1} = b$, $Z_{-1} = 0$, and $C_{-1} = 1$. Let N denote the number of iterations performed by the CORDIC algorithms. During each iteration i $(i = 0, 1, \cdots, N-1)$, the following computation is performed.

$$\begin{cases} X_i = X_{i-1} \\ Y_i = Y_{i-1} + d_i \cdot X_{i-1} \cdot C_{i-1} \\ Z_i = Z_{i-1} - d_i \cdot C_{i-1} \\ C_i = C_{i-1} \cdot 2^{-1} \end{cases} \tag{3.1}$$

where, $d_i = +1$ if $Y_i < 0$ and $d_i = -1$ otherwise. After N iterations of processing, we have $Z_N \approx -b/a$. Implementing this CORDIC algorithm using soft processors not only leads to compact designs but also offers dynamic adaptivity for practical application development. For example, many telecommunication systems have a wide dynamic data range, so it is desired that the number of iterations be dynamically adapted to the environment where the telecommunication systems are deployed. Also, for some CORDIC algorithms, the effective precision of the output data cannot be computed analytically. One example is the hyperbolic CORDIC algorithms. The effective output bit precision of these algorithms depends on the angular value Z_i during iteration i and needs to be determined dynamically.

• *Implementation*: The hardware architecture of our CORDIC algorithm for division based on MicroBlaze is shown in Figure 3.6. The customized hardware peripheral is configured with P processor elements (PEs). Each PE performs

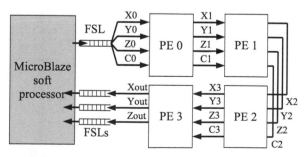

FIGURE 3.6: CORDIC algorithm for division with $P = 4$

TABLE 3.1:   Resource usage of the CORDIC based division and the block matrix multiplication applications as well as the simulation times using different simulation techniques

Designs	Estimated/actual resource usage		
	Slices	BRAMs	Multipliers
24 iteration CORDIC div. with $P = 2$	729 / 721	1 / 1	3 / 3
24 iteration CORDIC div. with $P = 4$	801 / 793	1 / 1	3 / 3
24 iteration CORDIC div. with $P = 6$	873 / 865	1 / 1	3 / 3
24 iteration CORDIC div. with $P = 8$	975 / 937	1 / 1	3 / 3
12×12 matrix mult. with 2×2 blocks	851 / 713	1 / 1	5 / 5
12×12 matrix mult. with 4×4 blocks	1043 / 867	1 / 1	7 / 7

Simulation time	
Our environment	ModelSim (Behavioral)
0.041 sec	35.5 sec
0.040 sec	34.0 sec
0.040 sec	33.5 sec
0.040 sec	33.0 sec
1.724 sec	1501 sec
0.787 sec	678 sec

one iteration of computation described in Equation 5.1. All the PEs form a linear pipeline. We consider 32-bit data precision in our designs. Since software programs are executed in a serial manner in the processor, only one FSL channel is used for sending the data from MicroBlaze to the customized hardware peripheral. The software program controls the number of iterations for each set of data based on the specific application requirement. To support more than 4 iterations for the configuration shown in Figure 3.6, the software program sends X_{out}, Y_{out} and Z_{out} generated by PE_3 back to PE_0 for further processing until the desired number of iterations is reached.

For the processing elements shown in Figure 3.6, C_0 is provided by the software program based on the number of times that the input data has passed through the linear pipeline. C_0 is sent out from the MicroBlaze processor to the FSL as a control word. That is, when there is data available in the corresponding FSL FIFO buffer and $Out\#\_control$ is high, PE_0 updates its local copy of C_0 and then continues to propagate it to the following PEs along the linear pipeline. For the other PEs, C_i is updated as $C_i = C_{i-1} \cdot 2^{-1}$, $i = 1, 2, \cdots, P-1$, and is obtained by right shifting C_{i-1} from the previous PE.

When performing division on a large set of data, the input data is divided into several sets. These sets are processed one by one. Within each set of data, the data samples are fed into the customized hardware peripheral consecutively in a back-to-back manner. The output data of the hardware peripheral is stored at the FIFO buffers of the data output FSLs and is sent back to the processor. The application designer needs to select a proper size for each set of data so that the results generated do not overflow the FIFO buffers of the data output FSL channels.

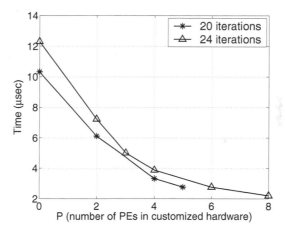

FIGURE 3.7: Time performance of the CORDIC algorithm for division ($P = 0$ denotes "pure" software implementations)

• *Design space exploration*: We consider different implementations of the CORDIC algorithm with different P, the number of processing elements used for implementing the linear pipeline. When more processing elements are employed in the design, the execution of the CORDIC division algorithm can be accelerated. However, the configuration of the MicroBlaze processor would also consume more hardware resources.

The time performance of various configurations of the CORDIC algorithm for division is shown in Figure 3.7, while its resource usage is shown in Table 3.1. The resource usage estimated using our design tool is calculated as shown in Section 3.4.3. The actual resource usage is obtained from the place-and-route reports (*.par* files) generated by ISE. For CORDIC algorithms with 24 iterations, attaching a customized linear pipeline of 4 PEs to the soft processor results in a 5.6x improvement in time performance compared with "pure" software implementation, while it requires 280 (30%) more slices.

• *Simulation speed-ups*: The simulation time of the CORDIC algorithm for division using our high-level co-simulation environment is shown Table 3.1. For the purpose of comparison, we also show the simulation time of the low-level behavioral simulation using ModelSim. For the ModelSim simulation, the time for generating the low-level implementations is not accounted for. We only consider the time needed for compiling the VHDL simulation models and performing the low-level simulation within ModelSim. Compared with the low-level simulation in ModelSim, our simulation environment achieves speed-ups in simulation time ranging from 825x to 866x and 845x on average for the four designs shown in Table 3.1.

3.5.1.2 Block Matrix Multiplication

In our design of block matrix multiplication, we first decompose the original matrices into a number of smaller matrix blocks. Then, the multiplication of these smaller matrix blocks is performed within the customized hardware peripheral. The software program is responsible for controlling data to and from the customized hardware peripheral, combining the multiplication results of these matrix blocks, and generating the resulting matrix. As is shown in Equation 3.2, to multiply two 4×4 matrices, A and B, we decompose them into four 2×2 matrix blocks respectively (i.e., $A_{i,j}$ and $B_{i,j}$, $1 \leq i,j \leq 2$). To minimize the required data transmission between the processor and the hardware peripheral, the matrix blocks of matrix A are loaded into the hardware peripheral column by column so that each block of matrix B only needs to be loaded once into the hardware peripheral.

$$
\begin{aligned}
A \cdot B &= \begin{pmatrix} A_{11} & A_{12} \\ A_{21} & A_{22} \end{pmatrix} \cdot \begin{pmatrix} B_{11} & B_{12} \\ B_{21} & B_{22} \end{pmatrix} \\
&= \begin{pmatrix} A_{11}B_{11} + A_{12}B_{21} & A_{11}B_{12} + A_{12}B_{22} \\ A_{21}B_{11} + A_{22}B_{21} & A_{21}B_{12} + A_{22}B_{22} \end{pmatrix}
\end{aligned}
\tag{3.2}
$$

• *Implementation*: The architecture of our block matrix multiplication based on 2×2 matrix blocks is shown in Figure 3.8. Similar to the design of the CORDIC algorithm, the data elements of matrix blocks from matrix B (e.g., b_{11}, b_{21}, b_{12} and b_{22} in Figure 3.8) are fed into the hardware peripheral as control words. That is, when data elements of matrix blocks from Matrix B are available in the FSL FIFO, $Out\#\_control$ becomes high and the hardware peripheral puts these data elements into the corresponding registers. Thus, when the data elements of matrix blocks from matrix A come in as normal data words, the multiplication and accumulation are performed accordingly to generate the output results.

• *Design space exploration*: We consider different implementations of the block matrix multiplication algorithm with different numbers of N, the size of the matrix blocks used by the customized hardware peripherals. For larger N employed by the customized hardware peripherals, a shorter execution time can potentially be achieved by the block matrix multiplication application. At the same time, more hardware resources would be used by the configuration of the MicroBlaze processor.

The time performance of various implementations of block matrix multiplication is shown in Figure 3.9 while their resource usage is shown in Table 3.1. For multiplication of two 12×12 matrices, the MicroBlaze processor with a customized hardware peripheral for performing 4×4 matrix block multiplication results in a 2.2x speed-up compared with "pure" software implementa-

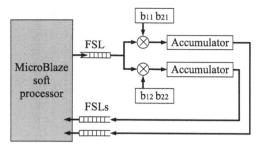

FIGURE 3.8: Matrix multiplication with customized hardware peripheral for matrix block multiplication with 2×2 blocks

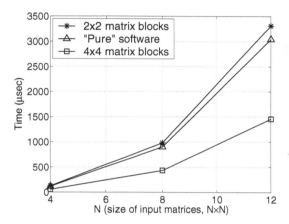

FIGURE 3.9: Time performance of our design of block matrix multiplication

tion. Also, attaching the customized hardware peripheral to the MicroBlaze processor requires an additional 767 (17%) more slices.

Note that attaching a customized hardware peripheral for computing 2×2 matrix blocks to the MicroBlaze processor results in worse performance for all the performance metrics considered. It uses 8.8% more execution time, 56 (8.6%) more slices and 2 (67%) more embedded multipliers compared with the corresponding "pure" software implementations. This is because in this configuration, the communication overhead for sending data to and back from the customized hardware peripheral is greater than the time saved by the parallel execution of multiplying the matrix blocks.

• *Simulation speed-ups*: Similar to Section 3.5.1.1, we compare the simulation time in the proposed cycle-accurate arithmetic level co-simulation environment with that of low-level behavioral simulation in ModelSim. Speed-ups in simulation time of 871x and 862x (866x on average) are achieved for the two different designs of the matrix multiplication application as shown in Table 3.1.

• *Analysis of simulation performance*: For both the CORDIC division application and the block matrix multiplication application, our co-simulation environment consistently achieves simulation speed-ups of more than 800x, compared with the low-level behavioral (functional) simulation using Model-Sim.

By utilizing the public C++ APIs (Application Program Interfaces) provided by the System Generator tool, we are able to tightly integrate the instruction set simulator for MicroBlaze processor with the simulation models for the other Simulink blocks. The simulators integrated into our co-simulation environment run in lock-step with each other. That is, the synchronization of the simulation processes within the hardware and software simulators and the exchange of simulation data between them occur at each Simulink simulation cycle. Thus, the hardware-software partitioning of the target application and the amount of data that needs to be exchanged between the hardware and software portions of the application would have little impact on the simulation speed-ups that can be achieved. Therefore, for the various settings of the two applications considered in this section, the variance of the simulation speed-ups is relatively small and we are able to obtain consistent speed-ups for all the design cases considered in our experiments.

3.5.2   Co-Simulation of a Complete Multi-Processor Platform

The co-simulation of the complete multi-processor platform is illustrated through the design of the 2-D DWT (Discrete Wavelet Transform) processing task of a motion JPEG2000 encoding application. Motion JPEG2000 encoding is a widely used image processing application. Performing JPEG2000 encoding on a 1024-pixel-by-768-pixel 24-bit color image takes around 8 seconds on a general purpose processor [31]. Encoding a 1 minute video clip with 50 frames per second would take over 6 hours. 2-D DWT is one of the most time-consuming processing tasks of the motion JPEG2000 encoding application. In the motion JPEG2000 application, the original input image is decomposed into a set of separate small image blocks. The 2-D DWT processing is performed on each of the small image blocks to generate output for each of them. The 2-D DWT processing within the motion JPEG2000 application exhibits a large degree of parallelism, which can be used to accelerate its execution. Employing a configurable multi-processor platform for 2-D DWT processing allows for rapid development while potentially leading to a significant speed-up in execution time.

The design of the configurable multi-processor platform for performing 2-D DWT processing is shown in Figure 3.10. Different numbers of MicroBlaze processors are used to concurrently process the input image data. Each of the processors has its local memory accessible through the instruction-side and data-side LMB buses. The local memory is used to store a copy of software programs and data for the MicroBlaze processors. By utilizing the dual-port

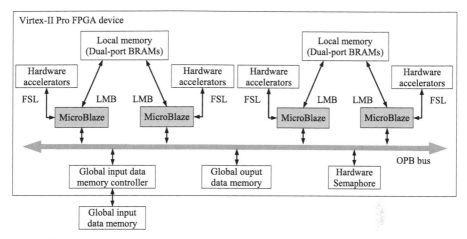

FIGURE 3.10: The configurable multi-processor platform with four MicroBlaze processors for the JPEG2000 encoding application

BRAM blocks available on Xilinx FPGAs, two processors share one BRAM block as their local memory. Customized hardware peripherals can be attached to the processors as hardware accelerators through the dedicated FSL interfaces. The multiple MicroBlaze processors are connected together using an OPB (On-chip Peripheral Bus) shared bus interface and get access to the global hardware resources (e.g., the off-chip global input data memory and the hardware semaphore).

A Single-Program-Multiple-Data (SPMD) programming model is employed for the development of software programs on the multi-processor platform. The input image data is divided into multiple small image blocks. Coordinated by an OPB based hardware semaphore, the multiple processors fetch the image blocks from the off-chip global input data memory through the OPB bus. The MicroBlaze processors store the input image blocks in their local memory. Then, the processors run the 2-D DWT software programs stored in their local memory to process the local copies of the image blocks. Once the 2-D DWT processing of an image block is finished, the processors send the output image data to the global input data memory coordinated by the hardware semaphore. See [50] for more details on the design of the multi-processing platform.

To simulate the multi-processor platform, multiple MicroBlaze Simulink blocks are used. Each of the MicroBlaze Simulink blocks is responsible for simulating one MicroBlaze processor. The 2-D DWT software programs are compiled and provided to the MicroBlaze Simulink blocks for simulation using the MicroBlaze instruction set simulator. The hardware accelerators are described within MATLAB/Simulink. Each processor and its hardware accelerators are placed in a MATLAB/Simulink subsystem. The arithmetic behavior of these MicroBlaze based subsystems can be verified separately by following the co-simulation procedure described in Section 3.5.1 before they are integrated into the complete multi-processor platform. The simulation of the OPB shared bus interface is performed by an OPB Simulink block. The global input/output data memory and the hardware semaphore are described within the MATLAB/Simulink modeling environment.

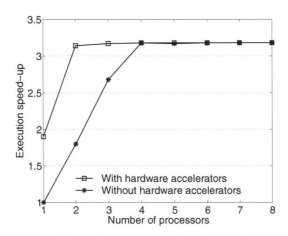

FIGURE 3.11:   Execution time speed-ups of the 2-D DWT task

• *Design space exploration*: We consider different configurations of the multi-processor platform as described above with the number of MicroBlaze processors used for processing the input image data.

The time performance of the multi-processor platform under different configurations for 2-D DWT processing is shown in Figure 3.11. For cases that do not employ hardware accelerators, the execution time speed-ups obtained from our arithmetic level co-simulation environment are consistent with those of the low-level implementations reported in [50]. For cases that either employ hardware accelerators or do not employ hardware accelerators, the maximum execution time speed-up achieved by the multi-processor platform is 3.18x. When no hardware accelerators are employed, the time performance of the multi-processor system fails to increase the number of MicroBlaze processors beyond 4. When hardware accelerators are employed, the time performance of the multi-processor system fails to increase the number of MicroBlaze processors beyond 2. Therefore, when no hardware accelerators are employed, the optimal configuration of the multi-processor platform is the configuration that uses 4 MicroBlaze processors. When hardware accelerators are employed, the optimal configuration of the multi-processor platform is the configuration that uses two MicroBlaze processors.

These optimal configurations of the multi-processor platform are identified within our arithmetic level co-simulation environment without the generation

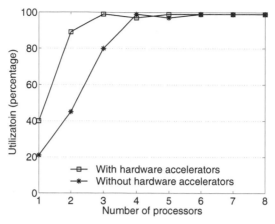

FIGURE 3.12: Utilization of the OPB bus interface when processing the 2-D DWT task

of low-level implementations and low-level simulation. This is compared against the development approach taken by James-Roxby et al. [50]. In [50], the optimal configurations of the multi-processor platform are identified after the low-level implementations of the multi-processor platform are generated and time-consuming low-level simulation is performed based on these low-level implementations.

Besides, the application designer can further identify the performance bottlenecks using the simulation information gathered during the arithmetic level co-simulation process. Considering the 2-D DWT application, the utilization of the OPB bus interface can be obtained from the arithmetic level co-simulation processes, which is shown in Figure 3.12. The OPB bus provides a shared channel for communication between the multiple processors. For the processing of one image block, each MicroBlaze processor needs to fetch the input data from the global data input memory through the OPB bus. After the 2-D DWT processing, each processor needs to send the result data to the global data memory through the OPB bus. The OPB bus thus acts as a processing bottleneck that limits the maximum speed-ups that can be achieved by the multi-processor platform.

• *Simulation speed-ups*: The simulation speed-ups achieved using our arithmetic level co-simulation approach as compared to those of low-level simulation using ModelSim are shown in Figure 3.13. Similar to Section 3.5.1, the time for generating the low-level simulation models that can be simulated in ModelSim is not accounted for for the experimental results shown in Figure 3.13. For the various configurations of the multi-processor platform, we are able to achieve simulation speed-ups ranging from 830x to 967x and 889x on average when simulating using our arithmetic level co-simulation environment as compared with the low-level behavioral simulation using ModelSim.

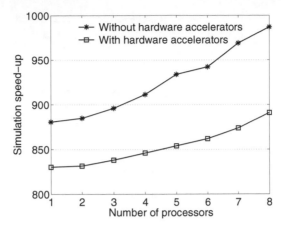

FIGURE 3.13:   Simulation speed-ups achieved by the arithmetic level
co-simulation environment

• *Analysis of simulation performance*: For the 2-D DWT processing task, the
proposed arithmetic level co-simulation environment is able to achieve simulation speed-ups very close to that of the two numerical processing applications
discussed in Section 3.5.1.

Higher simulation speed-ups are achieved as more processors are employed
in the designs. The MicroBlaze cycle-accurate instruction set simulator is a
manually optimized C simulation model. Simulating the execution of software
programs on MicroBlaze is much more efficient using the instruction set simulator than using the behavioral simulation model within ModelSim. As more
processors are employed in the design, a larger portion of the complete system
will be simulated using the instruction set simulators, which leads to increased
simulation speed-ups of the complete system.

In addition, with the same number of MicroBlaze processors employed in
the systems, simulation of systems without hardware accelerators consistently
have slightly higher simulation speed-ups compared to those with hardware
accelerators. This is mainly due to two reasons. One reason is the high simulation speed offered by the MicroBlaze instruction set simulator as discussed
above. Also, less Simulink blocks are used for describing the arithmetic-level
behavior of the systems when no hardware accelerators are used. This would
reduce the communication overhead between the Simulink simulation models
and the instruction-set simulation. This would contribute to the increase of
simulation speed-ups.

3.6   Summary

In this chapter, we propose a design space exploration technique based on arithmetic level cycle-accurate hardware-software co-simulation for application development using FPGA based configurable multi-processor platforms. An implementation of the proposed technique based on MATLAB/Simulink is provided to illustrate the construction of the proposed arithmetic level co-simulation environment. The design of several numerical computation and image processing applications were provided to demonstrate the effectiveness of the proposed design space exploration technique based on arithmetic level co-simulation.

Chapter 4

Energy Performance Modeling and Energy Efficient Mapping for a Class of Applications

4.1 Introduction

Reconfigurable hardware has evolved to become Reconfigurable System-on-Chips (RSoCs). Many modern reconfigurable hardware integrates general-purpose processor cores, reconfigurable logic, memory, etc., on a single chip. This is driven by the advantages of programmable design solutions over application specific integrated circuits and a recent trend in integrating configurable logic (e.g., FPGA, and programmable processors), offering the "best of both worlds" on a single chip.

In recent years, energy efficiency has become increasingly important in the design of various computation and communication systems. It is especially critical in battery operated embedded and wireless systems. RSoC architectures offer high efficiency with respect to time and energy performance. They have been shown to achieve energy reduction and increase in computational performance of at least one order of magnitude compared with traditional processors [11]. One important application of RSoCs is software defined radio (SDR). In SDR, dissimilar and complex wireless standards (e.g. GSM, IS-95, wideband CDMA) are processed in a single base station, where a large amount of data from the mobile terminals results in high computational requirement. The state-of-the-art RISC processors and DSPs are unable to meet the signal processing requirement of these base stations [23, 24]. Minimizing the power consumption has also become a key issue for these base stations due to their high computation requirements that dissipate a lot of energy as well as the inaccessible and distributed locations of the base stations. RSoCs stand out as an attractive option for implementing various functions of SDR due to their high performance, high energy efficiency, and reconfigurability.

In the systems discussed above, the application is decomposed into a number of tasks. Each task is mapped onto different components of the RSoC device for execution. By *synthesis*, we mean finding a mapping that determines an implementation for the tasks. We can map a task to hardware implementations on reconfigurable logic, or software implementations using the embedded processor core. Besides, RSoCs offer many control knobs (see Section 4.2 for details), which can be used to improve energy efficiency. In order to better exploit these control knobs, a performance model of the RSoC architectures and algorithms for mapping applications onto these architectures are required. The RSoC model should allow for a systematic abstraction of the available control knobs and enable system-level optimization. The mapping algorithms should capture the parameters from the RSoC model, the communication costs for moving data between different components on RSoCs, and the configuration costs for changing the configuration of the reconfigurable logic. These communication and configuration costs cannot be ignored compared with that required for computation. We show that a simple greedy mapping algorithm that maps each task onto either hardware or software, depending upon which dissipates the least amount of energy, does not always guarantee minimum energy dissipation in executing the application.

We propose a three-step design process to achieve energy efficient hardware/software co-synthesis on RSoCs. First, we develop a performance model that represents a general class of RSoC architectures. The model abstracts the various knobs that can be exploited for energy minimization during the synthesis process. Then, based on the RSoC model, we formulate a mapping problem for a class of applications that can be modeled as *linear pipelines*. Many embedded signal processing applications, such as the ones considered in this chapter, are composed of such a linear pipeline of processing tasks. Finally, a dynamic programming algorithm is proposed for solving the above mapping problem. The algorithm is shown to be able to find a mapping that achieves minimum energy dissipation in polynomial time.

We synthesize two beamforming applications onto Virtex-II Pro to demonstrate the effectiveness of our design methodology. Virtex-II Pro is a state-of-the-art RSoC device from Xilinx. In this device, PowerPC 405 processor(s), reconfigurable logic, and on-chip memory are tightly coupled through on-chip routing resources [110]. The beamforming applications considered can be used in embedded sonar systems to detect the direction of arrival (DOA) of close by objects. They can also be deployed at the base stations using the software defined radio technique to better exploit the limited radio spectrum [80].

The organization of this chapter is as follows. Section 4.2 identifies the knobs for energy-efficient designs on RSoC devices. Section 4.3 discusses related work. Section 4.4 describes the proposed RSoC model. Section 4.5 describes the class of *linear pipeline* applications we are targeting and formulates the energy-efficient mapping problem. Section 4.6 presents our dynamic programming algorithm. Section 4.7 illustrates the algorithm using two state-of-the-art beamforming applications. The modeling process and the energy dissipation results of implementing the two applications onto Virtex-II Pro are also given in this section. We conclude in Section 4.8.

4.2 Knobs for Energy-Efficient Designs

Various hardware and system level design knobs are available in RSoC architectures to optimize the energy efficiency of designs. For embedded processor cores, dynamic voltage scaling and dynamic frequency scaling can be used to lower the power consumption. The processor cores can be put into idle or sleep mode if desired to further reduce their power dissipation. For memory, the memory (SDRAM) on Triscend A7 CSoC devices can be changed to be in active, stand-by, disabled, or power-down state. Memory banking, which can be applied to the block-wise memory (BRAMs) in Virtex-II Pro, is another technique for low power designs. In this technique, the memory is split into banks and is selectively activated based on the use.

For reconfigurable logic, there are knobs at two levels that can be used to improve energy efficiency of the designs: low level and algorithm level.

Low-level knobs refer to knobs at the register-transfer or gate level. For example, Xilinx exposes the various features on their devices to designers through the *unisim* library [110]. One low-level knob is clock gating, which is employed to disable the clock to blocks to save power when the output of these blocks is not needed. In Virtex-II Pro, it can be realized by using primitives such as BUFGCE to dynamically drive a clock tree only when the corresponding block is used [110]. Choosing hardware *bindings* is another low-level knob. A binding is a mapping of a computation to a specific component on RSoC. Alternative realizations of a functionality using different components on RSoC result in different amounts of energy dissipation for the same computation. For example, there are three possible bindings for storage elements in Virtex-II Pro, which are registers, slice based RAMs, and embedded Block RAMs (BRAMs). The experiments by Choi et al. [18] show that registers and slice based RAMs have better energy efficiency for implementing small amounts of storage while BRAMs have better energy efficiency for implementing large amounts of storage.

Algorithm-level knobs refer to knobs that can be used during the algorithm development to reduce energy dissipation. It has been shown that energy performance can be improved significantly by optimizing a design at the algorithm level [78]. One algorithm-level knob is architecture selection. It plays a major role in determining the amount of interconnect and logic to be used in the design and thus affects the energy dissipation. For example, matrix multiplication can be implemented using a linear array or a 2-D array. A 2-D array uses more interconnects and can result in more energy dissipation compared with a 1-D array. Another algorithm-level knob is the algorithm selection. An application can be mapped onto reconfigurable logic in several ways by selecting different algorithms. For example, when implementing FFT, a radix-4 based algorithm would significantly reduce the number of complex multiplications that would otherwise be needed if a radix-2 algorithm is used.

TABLE 4.1:   Maximum operating frequencies of
different implementations of an 18×18-bit multiplication
on Virtex-II Pro

Design	VHDL(inferred)	VHDL(*unisim*)	IP cores
F_{max}	$\sim$120 MHz	$\sim$207 MHz	$\sim$354 MHz

Other algorithm-level knobs are parallel processing and pipelining.

As reconfigurable architectures are becoming domain-specific and integrate reconfigurable logic with a mix of resources, such as the ASMBL (Application Specific Modular Block) architecture proposed by Xilinx [86], more control knobs for application development are expected to be available on RSoCs in the future.

4.3   Related Work

Gupta and Wolf [37] and Xie [96] have considered the hardware/software co-design problem in the context of reconfigurable architectures. They use techniques, such as configuration pre-fetching, to minimize the execution time. Energy efficiency is not addressed by their research.

Experiments for re-mapping of critical software loops from a microprocessor to hardware implementations using configurable logic are carried out by Villarreal et al. [94]. Significant energy savings is achieved for a class of applications. However, a systematic technique that finds the optimal hardware and software implementations of these applications is not addressed. Such a systematic technique is a focus of this chapter.

A hardware-software bi-partitioning algorithm based on network flow techniques for dynamically reconfigurable systems has been proposed by Rakhmatov and Vrudhula [79]. While their algorithm can be used to minimize the energy dissipation, designs on RSoCs are more complicated than a hardware-software bi-partitioning problem due to the many control knobs discussed in the previous section.

A C-to-VHDL high-level synthesis framework is proposed by Gupta et al. [37]. The input to their design flow is C code and they employ a set of compiler transformations to optimize the resulting designs. However, generic HDL description is usually not enough to achieve the best performance as the recent FPGAs integrate many heterogeneous components. Use of device specific design constraint files and vendor IP cores as that in the MATLAB/Simulink based design flow plays an important role in achieving good performance. For example, Virtex-II Pro has embedded multipliers. We consider three implementations of an 18×18-bit multiplication using these multipliers. In the first implementation, we use VHDL for functional description. The use of embedded multipliers for implementing the functions is inferred by the synthesis tool. In the second implementation, this is accomplished by directly controlling the related low-level knobs through the *unisim* library. In the third implementation, we use the IP core for multiplication from Xilinx. Low-level knobs and the device specific design constraints are already applied for performance optimization during the generation of these IP cores. The maximum operating frequencies of these implementations are shown in Table 4.1. The implementation using IP core has by far the fastest maximum operating frequency. The reason for such performance improvement is that the specific locations of the embedded multipliers require appropriate connections between the multipliers and the registers around them. Use of appropriate location and timing constraints as in the generation of the IP cores leads to improved performance when using these multipliers [2]. It is expected that such constraint files and vendor IP cores will also have a significant impact on energy efficiency of the designs. Therefore, comparing with Gupta et al.'s approach, we consider task graphs as input to our design flow. In our approach, the energy efficiency of the designs is improved by making use of the various control knobs on the target device and parameterized implementations of the tasks.

System level tools are becoming available to synthesize applications onto architectures composed of both hardware and software components. Xilinx offers Embedded Development Kit (EDK) that integrates hardware and software development tools for Xilinx Virtex-II Pro [110]. In this design environment, the portion of the application to be synthesized onto software is described using C/C++ and is compiled using GNU *gcc*. The portion of the application to be executed in hardware is described using VHDL/Verilog and is compiled using Xilinx ISE. In the Celoxica DK2 tool [16], Handel-C (C with additional hardware description) is used for both hardware and software designs. Then, the Handel-C compiler synthesizes the hardware and software onto the device. While these system level tools provide high level languages to describe applications and map them onto processors and configurable hardware, none of them address synthesis of energy efficient designs.

4.4 Performance Modeling of RSoC Architectures

An abstraction of RSoC devices is proposed in this section. A model for Virtex-II Pro is developed to illustrate the modeling process.

4.4.1 RSoC Model

In Figure 4.1, the RSoC model consists of four components: a processor, a reconfigurable logic (RL) such as FPGA, a memory, and an interconnect. There are various implementations of the interconnect. For example, in Triscend CSoC [102], the interconnect between the ARM7 processor and the SDRAM is a local bus while the interconnect between the SDRAM and the configurable system logic is a dedicated data bus and a dedicated address bus. In Virtex-II Pro [110], the interconnect between the PowerPC processor and the RL is implemented using the on-chip routing resource. We abstract all these buses and connections as an interconnect with (possibly) different communication time and energy costs between different components. We assume that the memory is shared by the processor and the RL.

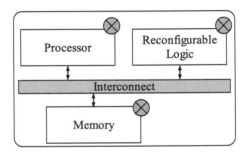

FIGURE 4.1:   The RSoC model

Since the operating state of the interconnect depends on the operating state of the other components, an operating state of the RSoC device, denoted as a *system state*, is thus only determined by the operating states for the processor, the RL, and the memory. Let S denote the set of all possible system states. Let $PS(s)$, $RS(s)$ and $MS(s)$ be functions of a system state s, $s \in S$. The output of these functions are integers that represent the operating states of the processor, the RL and the memory, respectively. An operating state of the processor corresponds to the state in which the processor is idle or is operating with a specific power consumption. Suppose that an idle mode and dynamic voltage scaling with $v - 1$ voltage settings are available on the processor. The processor is assumed to operate at a specific frequency for each of the voltage settings. Then, the processor has v operating states, $0 \leq PS(s) \leq v - 1$, with $PS(0) = 0$ being the state in which the processor is in the idle mode. The RL is idle when there is no input data and it is clock gated without switching activity on it. Thus, when the RL is loaded with a specific configuration, it can be in two different operating states depending on whether it is idle or processing the input data. Suppose that there are c configurations for the RL, then the RL has $2c$ operating states, $0 \leq RS(s) \leq 2c - 1$. We number the operating states of RL such that (a) for $0 \leq RS(s) \leq c - 1$, $RS(s)$ is the state in which the RL is idle, loaded with configuration $RS(s)$; (b) for $c \leq RS(s) \leq 2c - 1$, $RS(s)$ is the state which the RL is operating, loaded with configuration $RS(s) - c$. Each power state of the memory corresponds to an operating state. For example, when memory banking is used to selectively activate the memory banks, each combination of the activation states of the memory banks represents an operating state of the memory. Suppose that the memory has m operating states, then $0 \leq MS(s) \leq m - 1$. The operating state of the interconnect is related to the operating states of the other three components. Considering the above, the total number of distinct system states is $2vcm$.

The application is modeled as a collection of tasks with dependencies (see Section 4.5.1 for details). Suppose that task i' is to be executed immediately preceding task i. Also, suppose that task i' is executed in system state s' and task i is executed in system state s. If $s' \neq s$, a system state transition is required. The transition between different system states incurs a certain amount of energy. Our model consists of the following parameters:

- $\Delta EV_{PS(s'),PS(s)}$: state transition energy dissipation in the processor from $PS(s')$ to $PS(s)$

- $\Delta EC_{RS(s'),RS(s)}$: state transition energy dissipation in the RL from $RS(s')$ to $RS(s)$

- $\Delta EM_{MS(s'),MS(s)}$: memory state transition energy dissipation from $MS(s')$ to $MS(s)$

- IP: processor power consumption in the idle state
- IR: RL power consumption in the idle state
- $PM_{MS(s)}$: memory power consumption in state $MS(s)$
- $MP_{MS(s)}$: average energy dissipation for transferring one bit data between the memory and the processor when memory is in state $MS(s)$
- $MR_{MS(s)}$: average energy dissipation for transferring one bit data between the memory and the RL when memory is in state $MS(s)$

The system state transition costs depend not only on the source and destination system states of the transition but also on the requirement of the application. Let $\Delta_{i',i,s',s}$ be the energy dissipation for such system state transition. $\Delta_{i',i,s',s}$ can be calculated as

$$\Delta_{i',i,s',s} = \Delta EV_{PS(s'),PS(s)} + \Delta EC_{RS(s'),RS(s)} + \Delta EM_{MS(s'),MS(s)} + \Delta A_{i'i} \tag{4.1}$$

where, $\Delta A_{i'i}$ is the additional cost for transferring data from task i' to task i in a given mapping. For this given mapping, $\Delta A_{i'i}$ can be calculated based on the application models discussed in Section 4.5.1 and the communication costs $MP_{MS(s)}$ and $MR_{MS(s)}$.

4.4.2   A Performance Model for Virtex-II Pro

There are four components to be modeled in Virtex-II Pro. One is the embedded PowerPC core. Due to the limitations in measuring the effects of frequency scaling, we assume that the processor has only two operating states, *On* and *Off*, and is operating at a specific frequency when it is *On*. Thus, $v = 2$. We ignore IP since the PowerPC processor does not draw any power if it is not used in a design. $\Delta EV_{0,1}$ and $\Delta EV_{1,0}$ is also ignored since changing the processor states dissipates negligible amount of energy compared to when it performs computation. Two partial reconfiguration methods which are module based and small bit manipulation based [101] are available on the RL of Virtex-II Pro. For the small bit manipulation based partial reconfiguration, switching the configuration of one module on the device to another configuration requires downloading the difference between the configuration files for this module. This is different from the module based approach, which requires downloading the entire configuration file for the module. Thus, the small bit manipulation based partial reconfiguration has relatively low latency and energy dissipation compared with the module based one. Therefore, we use the small bit manipulation based method. We estimate the reconfiguration cost as the product of the number of slices used by the implementation and the average cost for downloading data for configuring one FPGA slice. According to [110], the energy for reconfiguring the entire Virtex-II Pro XC2VP20 device is calculated as follows. Let ICC_{Int} denote the current for powering the core of the device. We assume that the current for configuring the device is mainly drawn from ICC_{Int}. From the data sheet [110], $ICC_{Int} = 500$ mA@1.5V during configuration and $ICC_{Int} = 300$ mA@1.5V during normal operation, the reconfiguration power is estimated as $(500-300) \times 1.5 = 300$ mW. The time for reconfiguring the entire device using SelectMAP (50 MHz) is 20.54 ms. Thus, the energy for reconfiguring the entire device is 6162 μJ. There are 9280 slices on the device. Together with the slice usage from the post place-and-route report generated by the Xilinx ISE tool [110], we estimate the energy dissipation of reconfiguration as $\Delta EC_{RS(s'),RS(s)} = 6162 \times$ (total number of slices used by the RL in operating state $RS(s)$) / 9280 μJ. The quiescent power is the static power dissipated by the RL when it is on. This power cannot be optimized at the system level if we do not power on and off the RL. Thus, it is not considered in this chapter. Since IR represents the quiescent power, it is set to zero. We also ignore the energy dissipation for enabling/disabling clocks to the design blocks on the RL in the calculation of $\Delta EC_{RS(s'),RS(s)}$ since it is negligible compared with the other energy costs. For memory modeling, we use BRAM. It has only one available operating state, $m = 1$ and $MS(s) = 0$. Since the memory does not change its state, $\Delta EM_{MS(s'),MS(s)} = 0$. The BRAM dissipates negligible amount of energy when there is no memory access. We ignore this value so that $PM_0 = 0$. Using the power model from [18] and [115], energy dissipation MR_0 is estimated as 42.9 nJ/Kbyte. The communication between processor and memory fol-

FIGURE 4.2: A linear pipeline of tasks

lows certain protocols on the bus. This is specified by the vendor. Its energy efficiency is different depending on the bus protocols used. Energy cost MP_0 is measured through low-level simulation.

TABLE 4.2: Energy dissipation $E_{i,s}$ for executing task T_i in state s

	$0 \leq s \leq (v-1)cm - 1$	$(v-1)cm \leq s \leq vcm - 1$
1	$EP_{i,s}$	$ER_{i,s}$
2	$TP_{i,s} \cdot IR$	$TR_{i,s} \cdot IP$
3	$TP_{i,s} \cdot PM_{MS(s)}$	$TR_{i,s} \cdot PM_{MS(s)}$

	$vcm \leq s \leq (2v-1)cm - 1$
1	$EP_{i,s} + ER_{i,s}$
2	0
3	$\max(TP_{i,s}, TR_{i,s}) \cdot PM_{MS(s)}$

4.5 Problem Formulation

A model for a class of applications with linear dependency constraints is described in this section. Then, a mapping problem is formulated based on both the RSoC model and the application model.

4.5.1 Application Model

As shown in Figure 4.2, the application consists of a set of tasks, T_0, T_1, $T_2, \cdots, T_{n-1}$, with linear precedence constraints. T_i must be executed before initiating T_{i+1}, $i = 0, \cdots, n-2$. Due to the precedence constraints, only one task is executed at any time. The execution can be on the processor, on the RL, or on both. There is data transfer between adjacent tasks. The transfer can occur between the processor and the memory or between the RL and the memory, depending on where the tasks are executed.

The application model consists of the following parameters:

- D_{in}^i and D_{out}^i: amount of data input from memory to task T_i and data output from task T_i to memory.
- $EP_{i,s}$ and $TP_{i,s}$: processor energy and time cost for executing task T_i in system state s. $EP_{i,s} = TP_{i,s} = \infty$ if task T_i cannot be executed in system state s.
- $ER_{i,s}$ and $TR_{i,s}$: RL energy and time cost for executing task T_i in system state s. $ER_{i,s} = TR_{i,s} = \infty$ if task T_i cannot be executed in system state s.

4.5.2 Problem Definition

The energy efficient mapping problem is formulated based on the parameters of the RSoC model and the application model. We define an energy efficient mapping as the mapping that minimizes the overall energy dissipation for executing the application over all possible mappings.

During the execution of the application, a task can begin execution as soon as its predecessor task finishes execution. Thus, for any possible system state s, the processor and the RL cannot be in idle state at the same time. The total number of possible system states is $|S| = (2v - 1)cm$. Let the system states be numbered from 0 to $(2v-1)cm - 1$. Then, depending on the sources of energy dissipation, we divide the system states into three categories:

- For $0 \leq s \leq (v-1)cm - 1$, s denotes the system state in which the processor is in state $PS(s)$ $(1 \leq PS(s) \leq v - 1)$, the RL is in the idle state loaded with configuration $RS(s)$ and the memory is in state $MS(s)$. $PS(s)$, $RS(s)$ and $MS(s)$ are determined by solving equation $s = (PS(s) - 1)cm + RS(s)m + MS(s)$.
- For $(v-1)cm \leq s \leq vcm - 1$, s denotes the system state in which the processor is in the idle state $(PS(s) = 0)$, the RL is operating with configuration $RS(s) - c$ $(c \leq RS(s) \leq 2c - 1)$ and the memory is in state $MS(s)$. $PS(s)$, $RS(s)$ and $MS(s)$ are determined by solving equation $s = (RS(s) - c)m + MS(s) + (v - 1)cm$.
- For $vcm \leq s \leq (2v - 1)cm - 1$, s denotes the system state in which the processor is in state $PS(s)$ $(1 \leq PS(s) \leq v - 1)$, the RL is operating in state $RS(s) - c$ $(c \leq RS(s) \leq 2c - 1)$ and the memory is in state $MS(s)$. $PS(s)$, $RS(s)$ and $MS(s)$ are determined by solving equation $s = (PS(s) - 1)cm + (RS(s) - c)m + MS(s) + vcm$.

Let $E_{i,s}$ denote the energy dissipation for executing T_i in state s. Then, $E_{i,s}$ is calculated as the sum of the following:

- The energy dissipated by the processor and/or the RL that is/are executing T_i;
- If the processor or the RL is in the idle state, the idle energy dissipation of the component;
- The energy dissipated by the memory during the execution of T_i.

The above three sources of energy dissipation are calculated in Table 4.2.

We calculate the system state transition costs using Equation (4.1). Since a linear pipeline of tasks is considered, $i' = i - 1$. The energy dissipation for state transitions between the execution of two consecutive tasks T_{i-1} and T_i, namely, $\Delta_{i-1,i,s',s}$ is calculated as $\Delta EV_{PS(s'),PS(s)} + \Delta EC_{RS(s'),RS(s)} + \Delta EM_{MS(s'),MS(s)} + D_{out}^i \cdot MP_{MS(s')} + D_{in}^i \cdot MP_{MS(s)}$.

Let s_i denote the system state while executing T_i under a specific mapping, $0 \leq i \leq n - 1$. The overall system energy dissipation is given by

$$E_{total} = E_{0,s_0} + \sum_{i=1}^{n-1}(E_{i,s_i} + \Delta_{i-1,i,s'_{i-1},s_i}) \qquad (4.2)$$

Now, the problem can be stated as follows: *Find a mapping of tasks to system states, that is, a sequence of $s_0, s_1, \cdots, s_{n-1}$, such that the overall system energy dissipation given by Equation (4.2) is minimized.*

4.6   Algorithm for Energy Minimization

We create a trellis according to the RSoC model and the application model. Based on the trellis, a dynamic programming algorithm is presented in Section 4.6.2.

4.6.1   Trellis Creation

A trellis is created as illustrated in Figure 4.3. It consists of $n + 2$ steps, ranging from -1 to n. Each step corresponds to one column of nodes shown in the figure. Step -1 and step n consist of only one node 0, which represents the initial state and the final state of the system. Step i, $0 \leq i \leq n - 1$, consists of $|S|$ nodes, numbered from $0, 1, \cdots, |S| - 1$, each of which represents the system state for executing task T_i. The weight of node N_s in step i is the energy cost $E_{i,s}$ for executing task T_i in system state s. If task T_i cannot be executed in system state s, then $E_{i,s} = \infty$. Since node N_0 in step -1 and step n do not contain any tasks, $E_{-1,0} = E_{n,0} = 0$. There are directed edges (1) from node N_{-1} in step -1 to node N_j in step 0, $0 \leq j \leq |S| - 1$; (2) from node N_j in step $i - 1$ to node N_k in step i for $i = 1, \cdots, n - 1$, $0 \leq j, k \leq |S| - 1$; and (3) from node N_j in step $n - 1$ to node N_0 in step n, $0 \leq j \leq |S| - 1$. The weight of the edge from node $N_{s'}$ in step $i - 1$ to node N_s in step i is the system state transition energy cost $\Delta_{i-1,i,s',s}$, $0 \leq s', s \leq |S| - 1$. Note that all the weights are non negative.

4.6.2   A Dynamic Programming Algorithm

Based on the trellis, our dynamic programming algorithm is described below. We associate each node with a path cost $P_{i,s}$. Define $P_{i,s}$ as the minimum energy cost for executing $T_0, T_1, \cdots, T_i$ with T_i executed in node N_s in step i.

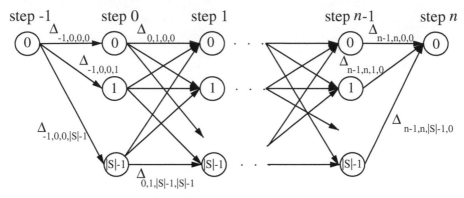

FIGURE 4.3:　The trellis

Initially, $P_{-1,0} = 0$. Then, for each successive step i, $0 \leq i \leq n$, we calculate the path cost for all the nodes in the step. The path cost $P_{i,s}$ for node N_s in step i is calculated as

• For $i = 0$:

$$P_{i,s} = \Delta_{-1,0,0,s} + E_{1,s}, \text{ for } 0 \leq s \leq |S| - 1 \qquad (4.3)$$

• For $1 \leq i \leq n - 1$:

$$P_{i,s} = \min_{0 \leq s' \leq |S|-1} \{P_{i-1,s'} + \Delta_{i-1,i,s',s} + E_{i,s}\}, \text{ for } 0 \leq s \leq |S| - 1 \qquad (4.4)$$

• For $i = n$:

$$P_{i,s} = \min_{0 \leq s' \leq |S|-1} \{P_{i-1,s'} + \Delta_{i-1,i,s',s} + E_{i,s}\}, \text{ for } s = 0 \qquad (4.5)$$

Only one path cost is associated with node N_0 in step n. A path that achieves this path cost is defined as a *surviving path*. Using this path, we identify a sequence of $s_0, s_1, \cdots, s_{n-1}$, which specifies how each task is mapped onto the RSoC device. From the above discussion, we have

THEOREM 4.1

The mapping identified by a surviving path achieves the minimum energy dissipation among all the mappings.

Since we need to consider $O((2v - 1)cm)$ possible paths for each node and there are $O((2v - 1)cm \cdot n)$ nodes in the trellis, the time complexity of the algorithm is $O(v^2 c^2 m^2 n)$. The configurations and the hardware resources are not reused between tasks in most cases, which means that the trellis constructed in Figure 4.3 is usually sparsely connected. Therefore, the following pre-processing can be applied to reduce the running time of the algorithm:

(1) nodes with ∞ weight and the edges incident on these nodes are deleted from the trellis; (2) the remaining nodes within each step are renumbered. After this two-step pre-processing, we form a *reduced trellis* and the dynamic programming algorithm is run on the reduced trellis.

4.7   Illustrative Examples

To demonstrate the effectiveness of our approach, we implement a broadband delay-and-sum beamforming application and an MVDR (minimum-variance distortionless response) beamforming application on Virtex-II Pro, a state-of-the-art reconfigurable SoC device. These applications are widely used in many embedded signal processing systems and in SDR [24].

4.7.1   Delay-and-Sum Beamforming

Using the model for Virtex-II Pro discussed in Section 4.4.2, implementing the delay-and-sum beamforming application is formulated as a mapping problem. This problem is then solved using the proposed dynamic programming algorithm.

4.7.1.1   Problem Formulation

The task graph of the broadband delay-and-sum beamforming application [42] is illustrated in Figure 4.4. A cluster of seven sensors samples data. Each set of the sensor data is processed by an FFT unit and then all the data is fed into the beamforming application. The output is the spatial spectrum response, which can be used to determine the directions of the objects near by. The application calculates twelve beams and is composed of three tasks with linear dependences: calculation of the relative delay for different beams according to the positions of the sensors (T_0), computation of the frequency responses (T_1), and calculation of the amplitude for each output frequency (T_2). The *data in* and *data out* are performed via the I/O pads on Virtex-II Pro. The number of FFT points in the input data depends on the frequency resolution requirements. The number of output frequency points is determined by the spectrum of interest. The three tasks can be executed either on the PowerPC processor or on the RL. The amount of data input (D_{in}^i) and output (D_{out}^i) varies with the tasks. For example, when both the numbers of FFT points and the output frequency points are 1024, D_{in}^1 and D_{out}^1 for task T_1 are 14 bytes and 84 Kbytes, respectively.

We employ the algorithm-level control knobs discussed in Section 4.2 to develop various designs on the RL. There are many possible designs. For the sake of illustration, we implement two designs for each task. One of the main

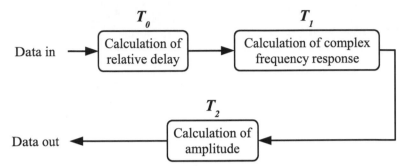

FIGURE 4.4:   Task graph of the delay-and-sum beamforming application

differences among these designs is the degree of parallelism, which affects the number of resources, such as I/O ports and *sine/cosine* look-up tables, used by the tasks. For example, one configuration of task T_0 handles two input data per clock cycle and requires more I/O ports than the other configuration that handles only one input per clock cycle. While the first configuration would dissipate more power and more reconfiguration energy than the second one, it reduces the latency to complete the computation. Similarly, one configuration for task T_2 uses two *sine/cosine* tables and thus can generate the output in one clock cycle while the other configuration uses only one *sine/cosine* table and thus requires two clock cycles in order to generate the output.

Each task is mapped on the RL to obtain $TR_{i,s}$ and $ER_{i,s}$ values. The designs for the RL were coded using VHDL and are synthesized using XST (Xilinx Synthesis Tool) provided by Xilinx ISE 5.2.03i [110]. The VHDL code for each task is parameterized according to the application requirements, such as the number of FFT points, and the architectural control knobs, such as precision of input data and hardware binding for storing intermediate data. The utilization of the device resources is obtained from the place-and-route report files (.par files). To obtain the power consumption of our designs, the VHDL code was synthesized using XST for XC2VP20 and the place-and-route design files (.ncd files) are obtained. Mentor Graphics ModelSim 5.7 was used to generate the simulation results (.vcd files). The .ncd and .vcd files were then provided to Xilinx XPower [112] to obtain the average power consumption. $TR_{i,s}$ is calculated based on our designs running at 50 MHz and 16-bit precision. $ER_{i,s}$ is calculated based on both $TR_{i,s}$ and power measurement from XPower.

For the PowerPC core on Virtex-II Pro, we developed C code for each task, compiled it using the *gcc* compiler for PowerPC, and generated the bitstream using the tools provided by Xilinx Embedded Development Kit (EDK). We used the SMART model from Synopsis [90], which is a cycle-accurate behavioral simulation model for PowerPC, to simulate the execution of the C code. The data to be computed is stored in the BRAMs of Virtex-II Pro. The latencies for executing the C code are obtained directly by simulating

TABLE 4.3: Energy dissipation of the tasks in the delay-and-sum beamforming application (μJ)

$E_{0,0}$	9.17	$E_{1,0}$	38.47	$E_{2,0}$	2.31
$E_{0,1}$	4.65	$E_{1,1}$	26.31	$E_{2,1}$	2.29
$E_{0,2}$	48.31	$E_{1,2}$	3039.52	$E_{2,2}$	123.24

the designs using ModelSim 5.7. The energy dissipation is obtained assuming a clock frequency of 300MHz and the analytical expression for processor power dissipation provided by Xilinx [110] as 0.9 mW/MHz × 300 MHz = 270 mW. Then, we estimate the $TP_{i,s}$ and $EP_{i,s}$ values. Note that the quiescent power is ignored in our experiments as discussed in Section 4.4.2.

Considering both the PowerPC and the FPGA, we have three system states for each of the three tasks on the reduced trellis after the pre-processing discussed in Section 4.6.2. Thus, $0 \leq s \leq 2$. Table 4.3 shows the $E_{i,s}$ values for the three tasks when the number of input FFT points and the output frequency points is 1024.

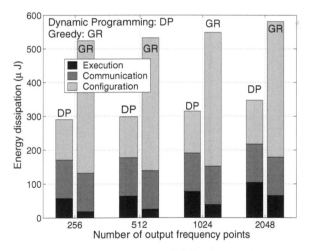

FIGURE 4.5: Energy dissipation of different implementations of the broadband delay-and-sum beamforming application (the input data is after 2048-point FFT processing)

For simple designs, the values of the parameters discussed above can be obtained through low-level simulations. However, for complex designs with many possible parameterizations, such low-level simulation can be time consuming. This is especially the case for designs on RL. However, using the domain-

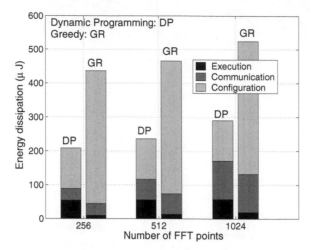

FIGURE 4.6:   Energy dissipation of different implementations of the broadband delay-and-sum beamforming application (the number of output frequency points is 256)

specific modeling technique proposed in [17] and the power estimation tool proposed by us in [69], it is possible to have rapid and fairly accurate system-wide energy estimation of data paths on RL without the time consuming low-level simulation.

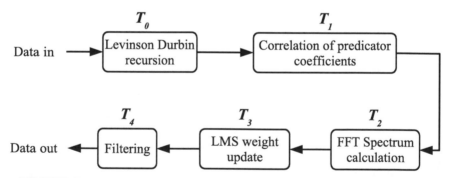

FIGURE 4.7:   Task graph of the MVDR beamforming application

4.7.1.2   Energy Minimization

We create a trellis with five steps to represent this beamforming application. After the pre-processing discussed in Section 4.6.2, step -1 and step 3 contain one node each while step 0, 1 and 2 contain three nodes each on the reduced

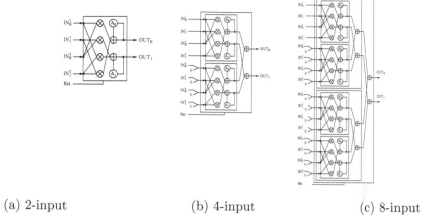

(a) 2-input (b) 4-input (c) 8-input

FIGURE 4.8: MAC architectures with various input sizes

trellis. By using the values described above, we obtain the weights of all the nodes and the edges in the trellis. Based on this, our dynamic programming based mapping algorithm is used to find the mapping that minimizes the overall energy dissipation.

For the purpose of comparison, we consider a greedy algorithm that always maps each task to the system state in which executing the task dissipates the least amount of energy. The results are shown in Figure 4.5 and Figure 4.6. For all the considered problem sizes, energy reduction ranging from 41% to 54% can be achieved by our dynamic programming algorithm over the greedy algorithm.

Considering the case where both the number of FFT points of the input data and the number of output frequency points are 2048, the greedy algorithm maps task T_0 on the RL. However, the dynamic programming algorithm maps this task on the processor and a 54% reduction of overall energy dissipation is achieved by doing so. The reason for the energy reduction is analyzed as follows. Task T_0 is executed efficiently on the RL for both the configuration files employed (ranging from 4.15 to 9.17 μJ). But the configuration costs for these two files are high (ranging from 272.49 to 343.03 μJ) since task T_0 needs *sine/cosine* functions. The Xilinx FPGA provides the CORE Generator lookup table [110] to implement the *sine/cosine* functions. For 16-bit input and 16-bit output *sine/cosine* lookup tables, the single output design (*sine* or *cosine*) needs 50 slices and the double output (both *sine* and *cosine*) design needs 99 slices. Two and three *sine/cosine* look-up tables are used in the two designs employed for T_0, which increases the reconfiguration costs for this task. The amount of computation energy dissipation of task T_0 is relatively small in this case and thus the configuration energy cost impacts the overall energy dissipation significantly. Therefore, executing the task on the processor dissipates less amount of energy than executing it on the RL.

4.7.2   MVDR Beamforming

Using a similar approach as in Section 4.7.1, we implemented an MVDR (Minimum Variance Distortionless Response) beamforming application on Virtex-II Pro. Details of the design process are discussed below.

4.7.2.1   Problem Formulation

The task graph of the MVDR beamforming application is illustrated in Figure 4.7. It can be decomposed into five tasks with linear constraints. In T_0, T_1 and T_2, we implemented a fast algorithm described in [42] for MVDR spectrum calculation. It consists of the Levinson Durbin recursion to calculate the coefficients of a prediction-error filter (T_0), correlation of the predictor coefficients (T_1), and the MVDR spectrum computation using FFT (T_2). This fast algorithm eliminates a lot of computation that is required by the direct calculation. We employ an LMS (Least Mean Square) algorithm (T_3) to update the weight coefficients of the filter due to its simplicity and numerical stability. A spatial filter (T_4) is used to filter the input data. The coefficients of the filter are determined by the previous tasks.

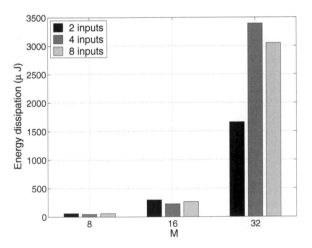

FIGURE 4.9:   Energy dissipation of task T_1 implemented using various MAC architectures

We considered the low-level and algorithm-level control knobs discussed in Section 4.2 and developed various designs for the tasks which are listed in Table 4.4. Different degrees of parallelism are employed in designs for task T_0 and T_1. Task T_2 uses FFT to calculate the MVDR spectrum. We employed the various FFT designs discussed in [18] which are based on the radix-4 algorithm as well as the design from Xilinx CORE Generator. Clock gating,

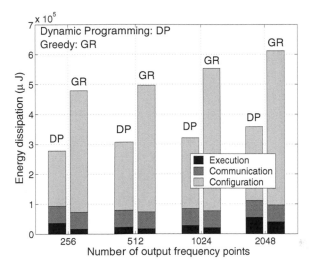

FIGURE 4.10: Energy dissipation of various implementations of the MVDR beamforming application ($M = 64$)

various degrees of parallelism, and memory bindings are used in these FFT designs to improve energy efficiency. V_p and H_p are the vertical and horizontal parallelism employed by the designs (see [18] for more details). Two different bindings, one using slice based RAMs and the other using BRAMs, were used to store the intermediate values. The number of dedicated multipliers used in the designs for task T_3 and T_4 is varied. Using the approach described in Section 4.7.1, we developed parameterized VHDL code for T_0, T_1, and T_2. Parameterized designs for T_3 and T_4 are realized using a MATLAB/Simulink based design tool developed by us in [69]. All the designs for the RL and the processor core were mapped on the corresponding components for execution. Values of the parameters of the RSoC model and the application model were obtained through low-level simulation. The synthesis designs for the RL run at a clock rate of 50 MHz. The data precision is 10 bits. Table 4.5 shows the $E_{i,s}$ values when $M = 8$ and the number of FFT points is 16 after the pre-processing discussed in Section 4.6.2.

Let M denote the number of antenna elements. For task T_0 and task T_1, we need to perform a complex multiply-and-accumulate (MAC) for problem sizes from 1 to M. Typically, when $M = 8, 16$ are used in the area of software defined radio while when $M = 32, 64$ are used in embedded sonar systems. There are several trade-offs which affect the energy efficiency when selecting the number of inputs to the complex MAC when implementing task T_0 and T_1. Architectures for complex MACs with 2, 4, and 8 inputs are shown in Figure 4.8. For a fixed M, using a complex MAC architecture handles more input data and at the same time reduces the execution latency. However, it dissipates more power. It also occupies more FPGA slices, which increases the

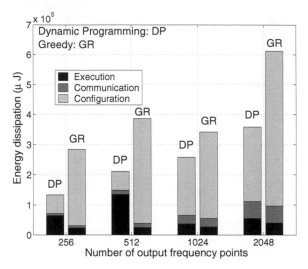

FIGURE 4.11: Energy dissipation of various implementations of the MVDR beamforming application (the number of points of FFT is 256)

configuration cost. The energy dissipation when using MACs with different input sizes for task T_1 is analyzed in Figure 4.9. While a MAC with input size of 4 is most energy efficient for task T_1 when $M = 8$, a MAC with input size of 2 is most energy efficient when $M = 64$. Also, the number of slices required for a complex MAC that can handle 2, 4, and 8 inputs are 100, 164, and 378, respectively. This incurs different configuration costs between the tasks.

4.7.2.2 Energy Minimization

A trellis with seven steps was created to represent the MVDR application. After applying the pre-processing technique discussed in Section 4.6.2, step -1 and step 5 contain one node each, step 0 and 1 contain four nodes each, step 2 contains seven nodes, step 3 contains 3 nodes, and step 4 contains 5 nodes on the reduced trellis. Using the values described in the previous section, we obtain the weights of all the nodes and the edges on the reduced trellis. The proposed dynamic programming algorithm is then used to find a mapping that minimizes the overall energy dissipation.

The results are shown in Figure 4.10 and Figure 4.11. For all the considered problem sizes, energy reduction from 41% to 46% are achieved by our dynamic programming algorithm over a greedy algorithm that maps each task onto either hardware or software, depending upon which dissipates the least amount of energy.

For both the dynamic programming algorithm and the greedy algorithm, task T_0 and task T_1 are mapped to the RL. However, the dynamic programming algorithm maps them to the design using 2-input complex MAC while

TABLE 4.4:   Various implementations of the
tasks on RL for the MVDR beamforming application

T_0 and T_1		T_2			
Design	No. of inputs to the MAC	Design	V_p	H_p	Binding
		1	1	2	SRAM
1	2	2	1	3	SRAM
2	4	3	1	3	BRAM
3	8	4	1	4	BRAM
		5	1	5	BRAM
		6	Xilinx design		

T_3		T_4	
Design	Number of embedded multipliers used	Design	Number of embedded multipliers used
1	11	1	32
2	7	2	28
3	3	3	24
		4	20
		5	16

TABLE 4.5:   Energy dissipation of the tasks in the MVDR
beamforming application (μJ)

$E_{0,0}$	145.2	$E_{1,0}$	61.0	$E_{2,0}$	67.2	$E_{3,0}$	21.4	$E_{4,0}$	5.0
$E_{0,1}$	161.9	$E_{1,1}$	47.3	$E_{2,1}$	432.3	$E_{3,1}$	69.9	$E_{4,1}$	10.1
$E_{0,2}$	274.4	$E_{1,2}$	59.8	$E_{2,2}$	363.2	$E_{3,2}$	82.1	$E_{4,2}$	17.7
$E_{0,3}$	24590.2	$E_{1,3}$	5397.8	$E_{2,3}$	2223.1			$E_{4,3}$	23.3
				$E_{2,4}$	12687.4			$E_{4,4}$	27.6
				$E_{2,5}$	167.7				
				$E_{2,6}$	16038.0				

the greedy algorithm maps them to the designs based on 4-input complex
MAC in cases such as when $M = 16$. Designs based on the 2-input complex
MAC are not the ones that dissipate the least amount of execution energy
for all the cases considered. However, the designs based on the MAC with
2 inputs occupy less amount of area than those based on the MACs with 4
and 8 inputs. By doing so, the reconfiguration energy during the execution
of the MVDR beamforming application is reduced by a range of 34% to 66%
in our experiments. For task T_2, the dynamic programming algorithm maps
it onto the PowerPC processor core while the greedy algorithm maps it on
the RL. While the parameterized FFT designs by Choi et al. [18] minimize
the execution energy dissipation of T_4 through the employment of parallelism,
radix, and choices of storage types, such energy minimization is achieved by
using more area on the RL. This increases the reconfiguration energy costs

and thus is not an energy efficient option when synthesizing the beamforming application.

4.8   Summary

A three-step design process for energy efficient application synthesis using RSoCs is proposed in this chapter. The design of two beamforming applications are presented to illustrate the effectiveness of our techniques.

While our work focused on energy minimization, with minor modifications, the performance model can be extended to capture throughput and area, and the mapping algorithm can be used to minimize the end-to-end latency. We used a linear task graph to model the application. When the task graph is generalized to be a DAG (Directed Acyclic Graph), it can be shown that the resulting optimization problem becomes NP-hard.

Chapter 5

High-Level Rapid Energy Estimation and Design Space Exploration

5.1 Introduction

The integration of hard processor cores along with the avalability of soft processor cores provide an exceptional design flexibility to reconfigure hardware. The Xilinx Virtex-II Pro/Virtex-4/Virtex-5 contain up to four embedded Power405 and Power440 hard processor IP cores. Examples of such soft processors include Nios from Altera [3], LEON3 from Gaisler [30], and MicroBlaze and PicoBlaze from Xilinx [97] among many other open-source or commercial soft processor implementations. Due to the superior performance and design flexibility, reconfigurable hardware has been used in the development of many embedded systems. As shown in Figure 5.1, for development using these modern FPGA devices, the application designer has the choice to map portions of the target applications to be executed either on soft processors as software programs or on customized hardware peripherals attached to the processors. While customized hardware peripherals are efficient for executing many data intensive computations, processors are efficient for executing many control and management functionalities as well as computations with tight data dependency between computation steps (e.g. recursive algorithms). In some design cases, using processors can lead to more compact designs and require much smaller amounts of resources than customized hardware peripherals. These compact designs may fit into a small FPGA device and thus, effectively reducing the quiescent energy dissipation [93].

Energy efficiency is an important performance metric. In the design of many embedded systems (e.g., software defined radio systems [24] and embedded digital signal processing applications [58]), energy efficiency is an important factor that determines the processing capability and life-span of these systems. Reconfigurable hardware makes it possible for an application to have many different hardware-software mappings and implementations. These different mappings and hardware-software implementations would result in significant variation in energy dissipation. Being able to rapidly obtain the energy dissipation of an application under a specific mapping and implementation is crucial to running these applications on FPGAs for energy efficiency. We address the

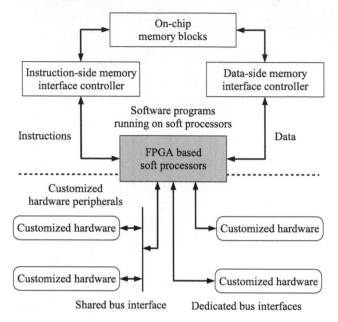

FIGURE 5.1:   FPGA-based hardware-software co-design

following design problem in this chaper.

   Problem definition: The FPGA device is configured with a soft processor
and several customized hardware peripherals. The processor and the hardware
peripherals communicate with each other through some specific bus protocols.
The target application is decomposed into a set of tasks. Each task can be
mapped on to the soft processor (software) or a specific customized hardware
peripheral (hardware) for execution. We are interested in a specific mapping
and execution scheduling of the tasks. Tasks executed on customized hardware
peripherals are described using high-level arithmetic modeling environments.
Examples of such environments include MATLAB/Simulink [60] and MILAN
[8]. Tasks executed on the soft processor are described as software programs
in C code. They are compiled using the C compiler and linker for the specific processor. One or more sets of sample input data are also given. Under
these assumptions, our objective is *to rapidly and (fairly) accurately obtain
the energy dissipation of the complete application.*

   There are two major challenges to achieving rapid and (fairly) accurate
energy estimation for hardware-software co-designs on FPGA platforms. One
challenge is that state-of-the-art design tools rely on time-consuming low-
level estimation techniques based on RTL (Register Transfer Level) and gate
level simulation models to obtain energy dissipation. While these low-level
energy estimation techniques can be accurate, they are too time-consuming
and would be intractable when used to evaluate energy performance of the different implementations on FPGAs, especially for software programs running

on soft processors. Taking the designs shown in Section 5.7 as an example, simulating ~2.78 msec actual execution time of a matrix multiplication application based on post place-and-route simulation models takes ~3 hours in ModelSim [63]. Using XPower [97] to analyze the simulation record file and calculate energy dissipation requires an additional ~1 hour. Thus, low-level energy estimation techniques are impractical for evaluating the energy performance of different implementations of the applications. Another challenge is that high-level energy performance modeling, which can provide rapid energy estimation, is difficult for designs using FPGAs. FPGAs have look-up tables as their basic elements. They lack a single high level model as that of general purpose processors, which can capture the energy dissipation behavior of all the possible implementations on them.

To address the above design problem, we propose a two-step rapid energy estimation technique for hardware-software co-design using FPGAs. In the first step, arithmetic level high-level abstractions are created for the hardware and software execution platforms. Based on the high-level abstractions, cycle-accurate arithmetic level hardware-software co-simulation is performed to obtain the arithmetic behavior of the complete system. Activity information of the corresponding low-level implementation of the system is estimated from the cycle-accurate arithmetic level co-simulation process. In the second step, by utilizing the estimated low-level activity information, an instruction-level energy estimation technique is employed to estimate the energy dissipation of software execution. Also, a domain-specific modeling technique is employed to estimate the energy dissipation of hardware execution. The energy dissipation of the complete system is obtained by summing up the energy dissipation for hardware and software execution.

To illustrate the proposed two-step energy estimation technique, we provide one implementation based on MATLAB/Simulink. We show the design of two widely used numerical computation applications to demonstrate the proposed technique. For the two numberical applications under consideration, our estimation technique achieves up to more than 6000 speed-ups compared with a state-of-the-art low-level simulation based energy estimation technique. Compared with the results from actual measurement, the energy estimates obtained using our approach achieve an average estimation error of 12% for various implementations of the two applications. The optimal implementations identified using our energy estimation technique achieve energy reductions up to 52% compared with other implementations considered in our experiments.

The organization of this chapter is described as follows. Section 5.2 discusses related work. Section 5.4 describes our rapid energy estimation technique. A MATLAB/Simulink implementation of our technique is also provided in this section. The design of two numerical computation applications is shown in Section 5.7 to demonstrate the effectiveness of our approach. Section 5.8 summarizes the chapter.

5.2   Related Work

5.2.1   Energy Estimation Techniques

The energy estimation techniques for designs using FPGAs can be roughly classified into two categories: low-level techniques and high-level techniques. The low-level energy estimation techniques utilize the low-level simulation results to obtain the switching activity information of the low-level hardware components (e.g., flip-flops and look-up tables). Switch activity is an important factor that determines the energy dissipation of a hardware component. The energy dissipation of each low-level hardware components are obtained using their switching activity information. The overall energy dissipation of an application is calculated as the sum of the energy dissipation of these low-level hardware components. The commercial tools such as Quartus II [3] and XPower [97], and the academic tools such as the one developed by Poon et al. [75] are based on low-level energy estimation techniques. Taking XPower as an example, the end user generates the low-level implementation of the complete design and creates a post place-and-route simulation model based on the low-level implementation. Then, sample input data is provided to perform post place-and-route simulation of the design using tools such as ModelSim [63]. A simulation file, which records the switching activity of each logic component and interconnect on the FPGA device, is generated during the low-level simulation. XPower calculates the power consumption of each logic component and interconnect using the recorded switching activity information and the pre-measured power characteristics (e.g., capacitance) of the low-level components and interconnects. The energy dissipation of a design is obtained by summing up the energy dissipation of the low-level hardware components and interconnects. While such low-level simulation based energy estimation techniques can be accurate, they are inefficient for estimating the energy dissipation of hardware-software co-designs using FPGAs. This is because modern FPGA devices contain multiple-million gates of configurable logic, and low-level post place-and-route simulations are very time consuming for such complicated devices. This is especially the case when using low-level simulation to simulate the execution of software programs running on the embedded processors.

One approach to perform high-level energy estimation is to first identify the parameters that have significant impact on energy dissipation and the energy models that estimate the power consumption of the applications. These parameters and energy models can be pre-defined through experiments, or provided by the application designer. During energy estimation, the designer provides the values of the parameters for their specific applications. These parameters are then applied to the energy models to retrieve the energy estimation of the target applications. The technique is used by tools such as the RHinO tool [25] and the web power analysis tools from Xilinx [103]. While energy estimation using this technique can be fast as it avoids the time-consuming low-level implementation and simulation process, its estimation accuracy can vary significantly among applications and application designers. One reason for that is because of the fact that different applications would demonstrate different energy dissipation behaviors. We show in [70] that using pre-defined parameters for energy estimation would result in energy estimation errors as high as 32% for input data with different characteristics. Another reason for the variance in energy dissipation is that requiring the application designer to provide these important parameters would demand him/her to have a deep understanding of the energy behavior of the target devices and the target applications, which can prove to be very difficult for many practical designs. This high-level energy estimation technique is especially impractical for estimating the energy estimation of software executions. A large number of different instructions with different energy dissipations are executed on soft processors for different software programs, which lead to different energy dissipation for a processor core.

Energy estimation for execution on processors has long been a research hot topic. There are a few prior works and attempts to estimate the energy dissipation of a few popular commercial and academic processors. For example, *JouleTrack* estimates the energy dissipation of software programs on StrongARM SA-1100 and Hitachi SH-4 processors [85]. *Wattch* [13] and *SimplePower* [116] estimate the energy dissipation of an academic processor. We proposed an instruction-level energy estimation technique in [71], which can provide rapid and fairly accurate energy estimation for soft processors. However, since these energy estimation frameworks and tools target a relative fixed processor execution platform, they do not address the energy dissipated by the customized hardware peripherals and the communication interfaces. Thus, they are not suitable for hardware-software co-designs on FPGA platforms.

5.2.2 High-Level Design Space Exploration

High-level design tools gain wider and wider adoption for application development. One important kind of high-level design tools are those based on high-level modeling environments, such as MATLAB/Simulink [60]. There are several advantages offered by these MATLAB/Simulink based design tools. One is that there is no need to know HDLs. This allows researchers and users from the signal processing community, who are usually familiar with the MATLAB/Simulink modeling environment, to get involved in the hardware design process. Another advantage is that the designer can make use of the powerful modeling environment offered by MATLAB/Simulink to perform arithmetic level simulation, which is much faster than behavioral and architectural simulations in traditional FPGA design flows [46].

There are several limitations when using the current MATLAB/Simulink design flow to perform design space exploration so as to optimize the energy performance of the applications. The state-of-the-art MATLAB/Simulink based design tools (e.g., *System Generator* [108] and DSP Builder [5]) have no support for *rapid* energy estimation for FPGA designs. One reason is that energy estimation using RTL (Register Transfer Level) simulation (which can be accurate) is too time consuming and can be overwhelming considering the fact that there are usually many possible implementations of an application on FPGAs. The other reason is that the basic elements of FPGAs are look-up tables (LUTs), which are too low-level an entity to be considered for high level modeling and rapid energy estimation. No single high level model can capture the energy dissipation behavior of all possible implementations on FPGAs. A rapid energy estimation technique based on domain-specific modeling is presented in [17] and is shown to be capable of quickly obtaining fairly accurate estimates of energy dissipation of FPGA designs. However, we are not aware of any tools that integrate such rapid energy estimation techniques.

Another limitation is that these tools do not provide interface for describing design constraints, traversing the MATLAB/Simulink design space, and identifying energy efficient FPGA implementations. Therefore, but algorithms such as the ones proposed in [68] are able to identify energy efficient designs for reconfigurable architectures, they cannot be directly integrated into the current MATLAB/Simulink based design tools.

To address the above limitations, we develop *PyGen*, an add-on programming interface that provides a textual scripting interface to MATLAB/Simulink based design tools. Our interface is written in Python scripting language [29], but it can be easily implemented using other scripting languages, such as the MATLAB M code, Tcl, etc. By creating an interface between Python and the MATLAB/Simulink based system level design tools, our tool allows the use of Python language for describing FPGA designs in MATLAB/Simulink. This provides several benefits. First, it enables the development of *parameterized* designs. Parameters related to application requirements (e.g., data precision) and those related to hardware implementations (e.g., hardware binding) can be captured by *PyGen* designs. It also enables *rapid* and *accurate* energy estimation by integrating a domain-specific modeling technique and using the switching activity information from MATLAB/Simulink simulation. Finally, it makes the identification of energy efficient designs possible by providing a flexible interface to traverse the design space.

5.3   Domain-Specific Modeling

As shown in Figure 5.2, domain-specific modeling is a hybrid (top-down followed by bottom-up) modeling technique. The modeling processing starts with a top-down analysis of the algorithms and the architectures for imple-

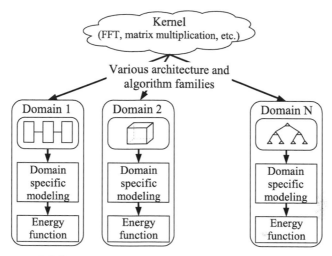

FIGURE 5.2:   Domain-specific modeling

menting the kernel. Through the top-down analysis, the various possible low-level implementations of the kernel are grouped into *domains* depending on the architectures and algorithms used. Through the grouping, we enforce a high-level architecture for the implementations that belong to the same domain. With such enforcement, it becomes possible to have a high-level model to capture the implementations within a domain. Analytical formulation of energy functions is derived within each domain to capture the energy behavior of the implementations belonging to the domain. Using the high-level energy models (i.e., the analytical energy functions), a bottom-up approach is followed to estimate the constants in these analytical energy functions for the identified domains through low-level sample implementations. This includes profiling individual system components through low-level simulations, real experiments, etc. These domain-specific energy functions are platform-specific. That is, the constants in the energy functions would have different values for different FPGA platforms. During the application development process, these energy functions can be used to rapidly estimate the energy dissipation of the hardware implementations belonging to their corresponding domains.

The domain-specific models can be hierarchical. For a kernel that consists of sub-kernels, the energy functions of the kernel can contain the energy functions of these sub-kernels. Besides, characteristics of the input data (e.g., switching activities) can have considerable impact on energy dissipation and are also inputs to the energy functions. This characteristic information is obtained through low-level simulation or through the arithmetic level co-simulation described in Section 5.4.1. More details about domain-specific energy performance modeling technique and the illustrative modeling process of a matrix multiplication example are presented as follows.

5.3.1   Domain-Specific Models for Matrix Multiplication

In this section, the development of an energy efficient matrix multiplication kernel is presented to illustrate the domain-specific modeling technique. And utilization the domain-specific models to optimize the energy performance of the kernel is explained.

Based on the location of the input and output matrices, we have two design scenarios: off-chip designs and on-chip designs. For off-chip design scenarios shown in Figure 5.3 (a), we assume that the input matrices are stored outside the FPGA. I/O ports are used for data access. While we assume that the input matrices are stored in an external memory outside the FPGAs, we do not include the energy used by the external memory. For the on-chip design scenario shown in Figure 5.3 (b), we store all input and output matrices in an on-chip memory of the FPGA devices. The on-chip memory refers to an embedded memory in FPGAs. For example, a block BRAM memory block in the Xilinx Virtex-II devices can be used for the on-chip memory. Thus, the energy used by the on-chip memory is included in the on-chip design.

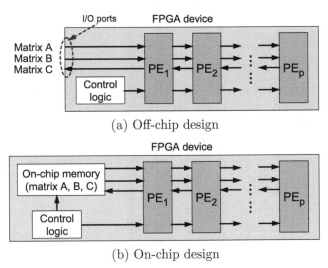

(a) Off-chip design

(b) On-chip design

FIGURE 5.3:   Design scenarios

We start the kernel development by analyzing the linear systolic matrix multiplication design proposed by Prasanna and Tsai in [76]. Their matrix multiplication design is proved to achieve the theoretical lower bound for linear systolic architecture in terms of latency. In addition, their linear pipeline design provides trade-offs between the number of registers and the latency. While their work focused on reducing the leading coefficient for the time complexity, our development focus is on minimizing energy dissipation under constraints for area and latency.

For performance analysis purposes, we implemented the latency-optimal systolic design in [76] on the Xilinx Virtex-II FPGA devices. The energy distribution profile of the design is shown in Figure 5.4. The profile reveals that much of the total energy is dissipated in the registers. For example, 78% of the energy is used in the registers for 12 × 12 matrix multiplication.

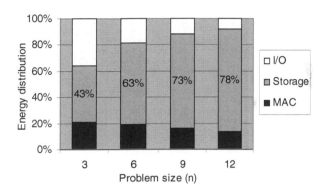

FIGURE 5.4:   Energy distribution of the design proposed in [76]

Once the energy hot spots are identified, energy-efficient algorithms and architectures for matrix multiplication are proposed. These algorithms and architectures are presented in the form of two theorems. Both of the two theorems employ a linear pipeline architecture consisting of a chain of processing elements (PEs). The first theorem (referred to as "Theorem 1" in the following) improves the latency-optimal algorithm for matrix multiplication [76] in terms of the number of registers used in the designs. The matrix multiplication designs based on Theorem 1 have optimal time complexity with a leading coefficient of 1 on a linear array. Theorem 1 can be parameterized to allow for trade-offs among energy dissipation, area, and latency. A domain-specific energy model is derived for these parameterized designs. Energy-efficient designs are identified using this domain-specific energy model under latency and area constraints. The second algorithm (referred to as "Theorem 2" in the following) is developed to exploit further increases in the density of FPGA devices to realize improvements in energy dissipation and latency. It uses more MACs and I/O ports compared with designs based on Theorem 1. The hardware architectures of the *PE*s of Theorem 1 and Theorem 2 are shown in Figure 5.5 and Figure 5.6. More details about the two theorems, including the pseudo-code for cycle-specific data movement, the detailed architectures, and a snapshot of an example computation can be found in [95].

A domain-specific performance model is applicable only to the design do-

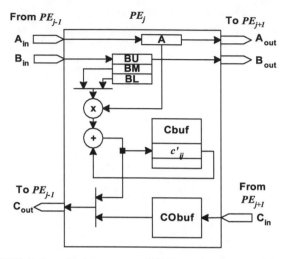

FIGURE 5.5:   Architecture of PE_j according to Theorem 1

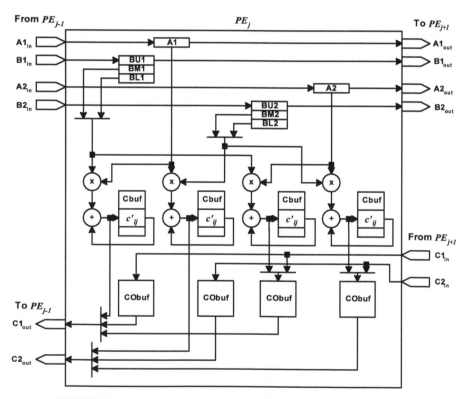

FIGURE 5.6:   Architecture of PE_j according to Theorem 2

main spanned by the family of algorithms and architectures being evaluated. The family represents a set of algorithm-architecture pairs that exhibit a common structure and similar data movement. The domain is a set of point designs resulting from unique combinations of algorithm and architecture level changes. The domain-specific energy model abstracts the energy dissipation to suit the design domain. The abstraction is independent of the commonly used levels such as gate, register, or system levels. Instead, it is based on the knowledge about the family of algorithms and architectures. The parameters are extracted considering their expected impact on the total energy performance. For example, if the number of MACs (multiply-and-accumulates) and the number of registers change values in a domain and are expected to be frequently accessed, a domain-specific energy model is built using these identified key parameters. The parameters may include elements at the gate, register, or system levels as needed by the domain. The domain-specific model is a knowledge-based model which exploits the kernel developer's knowledge about the algorithm and the architecture.

Using our knowledge of the two matrix multiplication domains, functions that represent energy dissipation, area and latency are derived. Beyond the simple complexity analysis, we make the functions as accurate as possible by incorporating implementation and target device details. For example, if the number of MACs is a key parameter, we implement a sample MAC on the target FPGA device to estimate its average power dissipation. Random input vectors, as many as are needed for the desired confidence interval [43], are generated for simulation. A power function representing the power dissipation as a function of m, the number of MACs, is generated. This power function is obtained for each module related to the key parameters. Based on the application developer's optimization goal and the time available for design, a balance needs to be struck between accuracy and simple representation of the functions. For the matrix multiplication designs developed using Theorem 1 and Theorem 2, the estimation error of the derived domain-specific energy and other performance functions ranges from 3.3% to 7.4%. This error range may be considered satisfactory, since the model is intended for algorithmic level analysis in the early stage of a design.

The two families of architectures and algorithms for matrix multiplication as specified by Theorem 1 and Theorem 2 form two domains. We limit algorithm-level exploration for energy optimization to the design space spanned by each of the two domains. For the two families of matrix multiplication architectures and algorithms shown in Figure 5.5 and Figure 5.6, the design parameters which represent the domain-specific design space under consideration, are shown in Table 5.1. Two parameters, n and r, are used in both theorems. In Theorem 1, n denotes the size of input matrices. r is introduced for block multiplication using sub-matrices of size $\frac{n}{r}$. In Theorem 2, r determines the number of I/O ports ($3r$), the number of MACs (r^2) and the sub-matrices of size $\frac{n}{r}$. Due to the nature of our algorithms, the number of each key module depends only on these two parameters.

For implementations on the target Virtex-II FPGA devices, we identify registers of 8-bit and 16-bit words, MACs, DRAMs (distributed memory blocks implemented using look-up tables), and BRAMs (embedded pre-compiled memory blocks) [99] to realize the key modules. Choosing specific values for the parameters in Table 5.1 results in a design point in the design space. For example, $n = 24$, $p = 6$, $reg = 4$, $m = 1$, $sram = 2$, $K_b = 2$, and $K_{io} = 0$ represents a design where 24×24 matrix multiplication is implemented using 6 PEs with 4 registers, one MAC, and two SRAMs per PE. The input and output matrices are stored in two ($\lceil 2 \times 24 \times 24/1024 \rceil = 2$) BSRAMs on the device and no I/O ports are used.

TABLE 5.1:   Range of parameters for Xilinx XC2V1500

Parameter	Range	FPGA constraints
Problem size (n)	2,3,4,...	
No. of PEs (p)	n/l, n is divisible by l, l is integer	
No. of registers/PE (reg)	$b^{2k+2}, b = 2, 3, 4, ...$ ($0 \le k \le \log_b n$)	8/16-bit registers
No. of MACs/PE (m)	b^{2k}	2-stage pipeline, embedded
No. of SRAMs/PE ($sram$)	$\lceil nb^k/16 \rceil$	16 words minimum
No. of BSRAMs/PE (K_b) (on-chip design)	$\lceil 2n^2/1024 \rceil$	1024 16-bit words minimum
No. of I/O ports (K_{io}) (off-chip design)	$3b^k$	8/16 bits

An energy model specific to the domain is constructed based on the analysis of the above key modules. It is assumed that, for each module of a given type (register, multiplier, SRAM, BRAM, or I/O port), the same amount of power is dissipated for the same input data, regardless of the locations of the modules on the FPGA device. This assumption significantly simplifies the derivation of system-wide energy dissipation functions. The energy dissipation for each module can be determined by counting the number of cycles the module stays in each power state and the low-level estimation of the power used by the module in the power state assuming average switching activity. Additional details of the model can be found in [17].

Table 5.2 lists the key parameters and the number of each key module in terms of the two parameters for each domain. In addition, it shows the latencies which also depend on the parameters. By choosing specific values

TABLE 5.2:   Number of modules and the latency of various designs

Domain	Parameters (range)	No. of PEs	No. of Registers/PE	No. of MACs/PE	No. of DRAMs/PE
Theorem 1	$n, r \ (r \leq n)$ n divisible by r	$\dfrac{n}{r}$	4	1	2
Theorem 2	$n, r \ (r \leq n)$ n divisible by r	$\dfrac{n}{r}$	$4r$	r^2	$2r^2$
Corollary 2	$n, m \ (1 \leq m \leq n^2)$ n^2 divisible by m	$\min\left(m, \dfrac{n^2}{m}\right)$	$\max\left(r, \dfrac{4m}{n}\right)$	$\max\left(1, \dfrac{m^2}{n^2}\right)$	$2\max\left(1, \dfrac{m^2}{n^2}\right)$

Domain	No. of BRAMs/PE	No. of I/O ports	Latency (cycles)
Theorem 1	$\left\lceil \dfrac{2n^2}{1024} \right\rceil$	3	$rn^2 + 2r^2 n$
Theorem 2	$\left\lceil \dfrac{2n^2}{1024} \right\rceil$	$3r$	$\dfrac{n^2}{r} + \dfrac{2n}{r}$
Corollary 2	$\left\lceil \dfrac{2n^2}{1024} \right\rceil$	$\max\left(3, \dfrac{3m}{n}\right)$	$\max\left(\dfrac{n^3}{m^3}, \dfrac{n}{m}\right)\min(n^2 + 2n, m^2 + 2m)$

for the parameters in Table 5.2, a different design is realized in the design space. For example, a design with $n = 16$ and $r = 4$ represents a design where 16×16 matrix multiplication is implemented using 4 PEs with 4 registers, one MAC, and one SRAM per PE.

5.3.2   High-Level Energy, Area, and Latency Functions

Using the key modules and the low-level analysis obtained from the above section, high-level functions that represent the energy dissipation, area, and latency are derived for both Theorem 1 and Theorem 2. The energy function of a design is approximated to be $\sum_i t_i P_i$, where t_i and P_i represent the number of active cycles and average power for module i. For example, P_{Mult} denotes the average power dissipation of the multiplier module. The average power is obtained from low-level power simulation of the module. The area function is given by $\sum_i A_i$, where A_i represents the area used by module i. In general, these simplified energy and area functions may not be able to capture all the implementation details needed for accurate estimation. However, we are concerned with algorithmic level comparisons, rather than accurate estimation. Moreover, our architectures are simple and have regular interconnections, and so the error between these functions and the actual values based on low-level simulation is expected to be small. In [95], we evaluate the accuracy of the energy and area functions. The latency function is obtained easily because the two theorems on which the designs are based already give us the latency in clock cycles for the different designs. It is observed that the estimation error of the derived functions ranges from 3.3% to 7.4% for energy dissipation and 3.3% to 4.1% for area. The estimation method (the domain-specific modeling) used in this paper and its accuracy are extensively discussed in [17].

Table 5.2 shows the number of modules used by the designs for $n \times n$ matrix multiplication with 8-bit input precision and 16-bit output precision. For the off-chip design, the I/O ports are used to fetch elements from outside the FPGA device. For the on-chip design, BRAMs of 1024 16-bit words are used for on-chip storage of the input matrices. SRAMs are configurable logic block based memory blocks used for storing intermediate results. The power and area values of each module are shown in Table 5.4. For example, P_{SRAM} is the average power used by SRAM (16-bit word), where x is the number of entries. In the actual implementation of a SRAM, the number of its entries should be multiples of 16. P_{offset} denotes the remaining power dissipation of a PE (after the modules have been accounted for) and takes care of glue logic and control logic. Similar numbers representing the area of each module are also obtained. A_{offset} denotes the area of a PE that accounts for glue logic and control logic. The latencies are obtained in terms of seconds by dividing them by the clock frequency. Using Table 5.2, functions that represent energy, area, and latency for Corollary 1, the parameterized formulation of Theorem 1, and Theorem 2 are shown in Table 5.3. Functions for other designs can be obtained in the same way. An average switching activity of 50% for input data to each module at a running frequency of 150MHz is assumed. Multiply operation is performed using dedicated embedded multipliers available in the Virtex-II device.

Note that throughput is important, since many applications for matrix multiplication process a stream of data. Our design in Corollary 1 is a pipelined architecture, with the first $\frac{n}{r}$ cycles of the computations on the next set of data being overlapped with the last $\frac{n}{r}$ cycles of the computations on the current set of data. Thus for a stream of matrices, an $\frac{n}{r} \times \frac{n}{r}$ sub-matrix can be processed every $\left(\frac{n}{r}\right)^2$ cycles. Thus the *effective latency* becomes $\left(\frac{n}{r}\right)^2$, which is the time between the arrivals of the first and last output data of the current computation. Hence, the design in Corollary 1 is a throughput-oriented design since one output is available every clock cycle for a stream of matrices. The design in Theorem 2 is also throughput-oriented since r output data items are available every clock cycle. Its effective latency becomes $\frac{n^2}{r}$.

5.3.3   Tradeoffs among Energy, Area, and Latency

The functions in Table 5.3 are used to identify tradeoffs among energy, area, and latency. For example, Figure 5.7 illustrates the tradeoffs among energy, area, and latency for 48×48 matrix multiplication for the off-chip and on-chip designs of Corollary 1. It can be used to choose energy-efficient designs to meet given area and latency constraints. For example, if 800 slices are available and

TABLE 5.3: Energy and time performance models

Corollary 1	
Metric	Performance model
Latency (cycles)	$L_{Cor1} = r^3 \left\{ (n/r)^2 + n/r \right\}$
Effective latency (cycles)	$L_{Cor1} = r^3 (n/r)^2$
Energy (on-chip)	$E_{Cor1} = L_{Cor1} \left\{ (n/r)(P_{Mult} + P_{Add} + 2P_{SRAM} + 4P_{R8} + 4P_{R16}) + \left\lceil 2n^2/1024 \right\rceil P_{BSRAM} + (n/r)P_{offset} \right\}$
Energy (off-chip)	$E_{Cor1} = L_{Cor1} \left\{ (n/r)(P_{Mult} + P_{Add} + 2P_{SRAM} + 4P_{R8} + 4P_{R16}) + 2P_I + P_O + (n/r)P_{offset} \right\}$
Area (on-chip)	$A_{Cor1} = (n/r)\left(A_{Mult} + A_{Add} + 2A_{SRAM} + 4A_{R8} + 4A_{R16} + A_{offset} \right)$ and $\left\lceil 2n^2/1024 \right\rceil$ BSRAMs
Area (off-chip)	$A_{Cor1} = (n/r)\left(A_{Mult} + A_{Add} + 2A_{SRAM} + 4A_{R8} + 4A_{R16} + A_{offset} \right)$ and two 8-bit input ports, one 16-bit output port

Theorem 2	
Metric	Performance model
Latency (cycles)	$L_{Thm2} = n^2/r + 2n/r$
Effective latency (cycles)	$L_{Thm2} = n^2/r$
Energy (on-chip)	$E_{Thm2} = L_{Thm2} \left\{ nr(P_{Mult} + P_{Add} + 2P_{SRAM}) + n(4P_{R8} + 4P_{R16}) + \left\lceil 2n^2/1024 \right\rceil P_{BSRAM} + (n/r)P_{offset} \right\}$
Energy (off-chip)	$E_{Thm2} = L_{Thm2} \left\{ nr(P_{Mult} + P_{Add} + 2P_{SRAM}) + n(4P_{R8} + 4P_{R16}) + 2rP_I + rP_O + (n/r)P_{offset} \right\}$
Area (on-chip)	$A_{Thm2} = nr\left(A_{Mult} + A_{Add} + 2A_{SRAM} \right) + n(4A_{R8} + 4A_{R16}) + (n/r)A_{offset}$ and $\left\lceil 2n^2/1024 \right\rceil$ BSRAMs
Area (off-chip)	$A_{Thm2} = nr\left(A_{Mult} + A_{Add} + 2A_{SRAM} \right) + n(4A_{R8} + 4A_{R16}) + (n/r)A_{offset}$ and $2r$ 8-bit input ports, r 16-bit output ports

the latency should be less than 6,000 cycles ($36\mu s$), an energy-efficient design is obtained using $\frac{n}{r} = 4$. The energy dissipation, area, and latency for such a design, evaluated using the functions in Table 5.3, are $6.85\mu J$, 524 slices, and 5400 cycles ($32.4\mu s$), respectively. Figure 5.7 shows that as the block size ($\frac{n}{r}$) increases, the area increases and the latency decreases, because the degree of parallelism increases. While the energy dissipation decreases until $\frac{n}{r} = 15$ or 16, it starts increasing afterwards. The reason for this behavior is as follows. The energy used by the local storages, Cbuf and CObuf, is $2n^3(0.126\lceil\frac{1}{16}\frac{n}{r}\rceil + 2.18)$ and is hence proportional to $O(\frac{n^4}{r})$. The energy used by the rest of modules, except I/O, are proportional to $O(n^3)$. The energy for I/O is proportional to $O(rn^2)$. As $\frac{n}{r}$ increases (r decreases), the energy used by I/O decreases relatively faster and thus the total energy decreases. However, after $\frac{n}{r} > 16$, the energy used by the local storage becomes the dominant factor. This helps us to identify the optimal block size for energy-efficient matrix multiplication.

TABLE 5.4:   Power and area functions for various modules

Module	Power Function (mW)	Area Function (slice)
Block multiplier (8x8 bit)	$P_{Mult} = 12.50$	$A_{Mult} = 16*$
Adder (8 bit)	$P_{Add} = 2.77$	$A_{Add} = 4$
SRAM (16-bit word, x number of entries)	$P_{SRAM} = 0.126\left\lceil\dfrac{x}{16}\right\rceil + 2.18$	$A_{SRAM} = 18.44\left\lceil\dfrac{x}{16}\right\rceil + 16.40$
BSRAM (16 bit, 1024 entries)	$P_{BSRAM} = 16.37$	$A_{BSRAM} = 16*$
Register (8 bit)	$P_{R8} = 2.12$	$A_{R8} = 4$
Register (16 bit)	$P_{R16} = 2P_{R8}$	$A_{R16} = 8$
Output port (16 bit)	$P_O = 70$	
Input port (8 bit)	$P_I = 10$	

* Block multiplier or BSRAM uses area equivalent to 16 slices.

Tradeoff analysis for the on-chip model also shows similar behavior. The on-chip design uses BRAMs instead of I/O ports. Since the energy used in I/O ports is more than the energy used in the BSRAMs, the energy used in the on-chip design is less than the energy used in the off-chip design. However, the choice between the off-chip and on-chip design depends on whether the matrix multiplication is stand-alone or a part of an application (e.g. an application consists of multiple kernels).

Theorem 2 provides asymptotic improvement in energy and latency performance in the on-chip model. As shown in Table 5.2, asymptotically, the energy dissipated in the BSRAMs and the latency of the Xilinx reference design increase as $O(n^5)$ and $O(n^3)$, respectively, assuming a unit of energy is

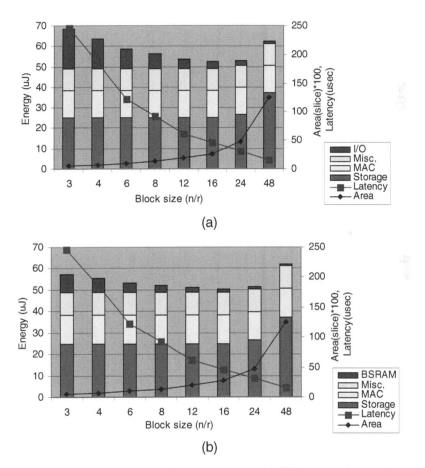

(a)

(b)

FIGURE 5.7: Energy, area, latency trade-offs of Theorem 1 as a function of the block size (n/r), (a) off-chip design and (b) on-chip design for $n = 48$

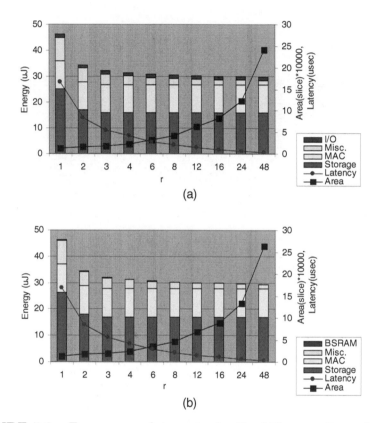

FIGURE 5.8:   Energy, area, latency trade-offs of Theorem 2 as a function of r, (a) off-chip design and (b) on-chip design for $n = 48$

used per cycle for retaining a word in the BSRAM. Energy dissipation and latency for the designs based on Theorem 2 and [76] increase as $O(n^4)$ and $O(n^2)$, respectively, under the same assumptions. Theorem 2 improves these complexities to $O(\frac{n^4}{r})$ and $O(\frac{n^2}{r})$, respectively, where $\frac{n}{r}$ is the block size for block multiplication and n is divisible by r with $r \geq 1$. Further increases in the density of FPGAs can be used to increase the number of multipliers and hence nr, leading to asymptotic reduction in energy dissipation and latency.

Figure 5.8 shows the trade-offs among energy, area, and latency for Theorem 2. As the value of r increases (or block size $\frac{n}{r}$ decreases), the area increases and the latency decreases. However, the energy dissipation continuously decreases. Thus, the designs based on Theorem 2 reach the minimal point of energy dissipation when the block size is the smallest unlike the designs based on Corollary 1. Note that the local storages consists of registers, Cbufs, and CObufs. The energy used by the registers in the designs based on Theorem 2 is $O(\frac{n^3}{r})$ while the energy used by the registers in the designs based on Corollary 1 is $O(n^3)$ for the same problem size. Thus, the energy used by the registers in the designs based on Theorem 2 decreases as r increases while the energy used by the registers in the designs based on Corollary 1 is constant. The same analysis applies to the energy complexity of BRAMs.

5.4   A Two-Step Rapid Energy Estimation Technique

We propose a two-step approach for rapid energy estimation of hardware-software co-design using FPGAs, which is illustrated in Figure 5.9. In the first step, we build a cycle-accurate arithmetic level hardware-software co-simulation environment to simulate the applications running on FPGAs. By "arithmetic level", we denote that only the arithmetic aspects of the hardware-software execution are captured by the co-simulation environment. For example, low-level implementations of multiplication on Xilinx Virtex-II FPGAs can be realized using either slice-based multipliers or embedded multipliers. The arithmetic level co-simulation environment only captures the multiplication arithmetic aspect during the simulation process. By "cycle-accurate", we denote that for each of the clock cycles under simulation, the arithmetic level behavior of the complete FPGA hardware platform predicted by the cycle-accurate co-simulation environment should match with the arithmetic level behavior of the corresponding low-level implementations. When simulating the execution of software programs on soft processors, the cycle-accurate co-simulation should take into account the number of clock cycles required for completing a specific instruction (e.g., the multiplication instruction of the MicroBlaze processor takes three clock cycles to finish) and the processing pipeline of the processor. Also, when simulating the execution on cus-

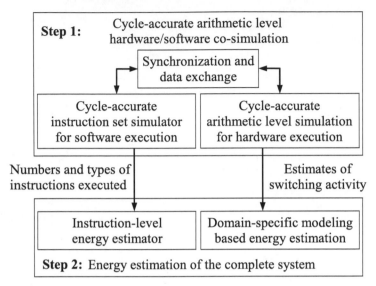

FIGURE 5.9:   The two-step energy estimation approach

tomized hardware peripherals, the cycle-accurate co-simulation should take into account the delay in number of clock cycles caused by the processing pipelines within the customized hardware peripherals. Our arithmetic level simulation environment ignores the low-level implementation details and focuses on only the arithmetic behavior of the designs. It can greatly speed up the hardware-software co-simulation process. In addition, the cycle-accurate property is maintained among the hardware and software simulators during the co-simulation process. Thus, the activity information of the corresponding low-level implementations of the hardware-software execution platform, which are used in the second step for energy estimation, can be accurately estimated from the high-level co-simulation process.

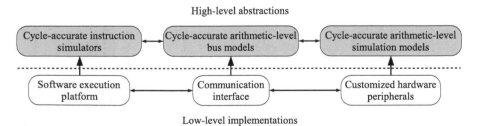

FIGURE 5.10:   Software architecture of our hardware-software
co-simulation environment

In the second step, we utilize the information gathered during the arithmetic level co-simulation process for rapid energy estimation. The types and the number of each type of instructions executed on soft processors are obtained from the cycle-accurate instruction simulation process. By utilizing an instruction-level energy estimation technique, the instruction execution information is used to estimate the energy dissipation of the software programs running on the soft processor. For customized hardware implementations, the switching activities of the low-level implementations are estimated by analyzing the switching activities of the arithmetic level simulation results. Then, with the estimated switching activity information, energy dissipation of the hardware peripherals is estimated by utilizing a domain-specific energy performance modeling technique proposed by Choi et al. [17]. Energy dissipation of the complete systems is obtained by summing up the energy dissipation of the software programs and the hardware peripherals.

For illustrative purposes, an implementation of our rapid energy estimation technique based on MATLAB/Simulink is described in the following subsections.

5.4.1   Step 1: Cycle-Accurate Arithmetic Level Co-Simulation

The software architecture of an arithmetic level hardware-software co-simulation environment for designs using FPGAs is illustrated in Figure 5.10. The low-level implementation of the FPGA based hardware-software execution platform consists of three major components: the *soft processor* for executing software programs; *customized hardware peripherals* as hardware accelerators for parallel execution of some specific computations; and *communication interfaces* for exchanging data and control signals between the processor and the customized hardware components. An arithmetic level ("high-level") abstraction is created for each of the three major components. These high-level abstractions are simulated using their corresponding simulators. These hardware and software simulators are tightly integrated into our co-simulation environment and concurrently simulate the arithmetic behavior of the hardware-software execution platform. Most importantly, the simulation among the integrated simulators are synchronized at each clock cycle and provide cycle accurate simulation results for the complete hardware-software execution platform. Once the design process using the arithmetic level abstraction is completed, the application designer specifies the required low-level hardware bindings for the arithmetic operations (e.g., binding the embedded multipliers for realization of the multiplication arithmetic operation). Finally, low-level implementations of the complete platform with corresponding arithmetic behavior can be automatically generated based on the arithmetic level abstractions of the hardware-software execution platforms.

We provide an implementation of the arithmetic level co-simulation approach based on MATLAB/Simulink, the software architecture of which is shown in Figure 5.11. The four major functionalities of our MAT-

LAB/Simulink based co-simulation environment are described in the following paragraphs.

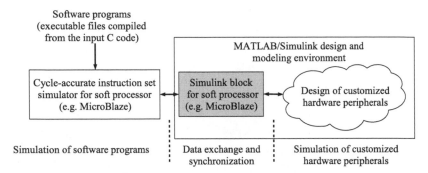

FIGURE 5.11:   An implementation of the hardware-software co-simulation environment based on MATLAB/Simulink

- *Cycle-accurate simulation of the software programs*: The input C programs are compiled using the compiler for the specific processor (e.g., the GNU C compiler *mb-gcc* for MicroBlaze) and translated into binary executable files (e.g., *.ELF* files for MicroBlaze). These binary executable files are then simulated using a cycle-accurate instruction set simulator for the specific processor. Taking the MicroBlaze processor as an example, the executable *.ELF* files are loaded into *mb-gdb*, the GNU C debugger for MicroBlaze. A cycle-accurate instruction set simulator for the MicroBlaze processor is provided by Xilinx. *mb-gdb* sends instructions of the loaded executable files to the MicroBlaze instruction set simulator and perform cycle-accurate simulation of the execution of the software programs. *mb-gdb* also sends/receives commands and data to/from MATLAB/Simulink through the Simulink block for the soft processor and interactively simulates the execution of the software programs in concurrence with the simulation of the hardware designs within MATLAB/Simulink.
- *Simulation of customized hardware peripherals*: The customized hardware peripherals are described using the MATLAB/Simulink based FPGA design tools. For example, *System Generator* supplies a set of dedicated Simulink blocks for describing parallel hardware designs using FPGAs. These Simulink blocks provide arithmetic level abstractions of the low-level hardware components. There are blocks that represent the basic hardware resources (e.g., flip-flop based registers and multiplexers), blocks that represent control logic, mathematical functions, and memory, and blocks that represent proprietary IP (Intellectual Property) cores (e.g., the IP cores for Fast Fourier Transform and finite impulse filters). Considering the *Mult* Simulink block for multiplication provided by *System Generator*, it captures the arithmetic behavior of multiplication by presenting at its output port the product of the values

presented at its two input ports. The low-level design trade-off of the *Mult* Simulink block that can be realized using either embedded or slice-based multipliers is not captured in its arithmetic level abstraction. The application designer assembles the customized hardware peripherals by dragging and dropping the blocks from the block set to his/her designs and connecting them via the Simulink graphic interface. Simulation of the customized hardware peripherals is performed within the MATLAB/Simulink. MATLAB/Simulink maintains a simulation timer for keeping track of the simulation process. Each unit of simulation time counted by the simulation timer equals to one clock cycle experienced by the corresponding low-level implementations. Finally, once the design process completes, the low-level implementations of the customized hardware peripherals are automatically generated by the MATLAB/Simulink based design tools.

• *Data exchange and synchronization among the simulators*: The Simulink block for soft processor is responsible for exchanging simulation data between the software and hardware simulators during the co-simulation process. MATLAB/Simulink provides *Gateway In* and *Gateway Out* Simulink blocks for separating the simulation of the hardware designs described using *System Generator* with the simulation of other Simulink blocks (including the MicroBlaze Simulink blocks). These *Gateway In* and *Gateway Out* blocks identify the input/output communication interfaces of the customized hardware peripherals. For the MicroBlaze processor, the MicroBlaze Simulink block sends the values of the processor registers stored at the MicroBlaze instruction set simulator to the *Gateway In* blocks as input data to the hardware peripherals. Vice versa, the MicroBlaze Simulink blocks collect the simulation output of the hardware peripherals from *Gateway Out* blocks and use the output data to update the values of the processor registers stored at the MicroBlaze instruction set simulator. The Simulink block for soft processor also simulates the communication interfaces between the soft processor and the customized hardware peripherals described in MATLAB/Simulink. For example, the MicroBlaze Simulink block simulates the communication protocol and the FIFO buffers for communication through the Xilinx dedicated FSL (Fast Simplex Link) interfaces [97].

Besides, the Simulink block for soft processor maintains a global simulation timer which keeps track of the simulation time experienced by the hardware and software simulators. When exchanging the simulation data among the simulators, the Simulink block for soft processor takes into account the number of clock cycles required by the processor and the customized hardware peripherals to process the input data as well as the delays caused by transmitting the data between them. Then, the Simulink block increases the global simulation timer accordingly. By doing so, the hardware and software simulation are synchronized on a cycle accurate basis.

5.4.2   Step 2: Energy Estimation

The energy dissipation of the complete system is obtained by summing up the energy dissipation of software execution and that of hardware execution, which are estimated separately by utilizing the activity information gathered during the arithmetic level co-simulation process.

5.4.2.1   Instruction-Level Energy Estimation for Software Execution

An instruction-level energy estimation technique is employed to estimate the energy dissipation of the software execution on the soft processor. An instruction energy look-up table is created which stores the energy dissipation of each type of instructions for the specific soft processor. The types and the number of each type of instructions executed when the software program is running on the soft processor are obtained during the arithmetic level hardware-software co-simulation process. By querying the instruction energy look-up table, the energy dissipation of these instructions is obtained. Energy dissipation for executing the software program is calculated as the sum of the energy dissipation of the instructions.

We use the MicroBlaze processor to illustrate the creation of the instruction energy look-up table. The overall flow for generating the look-up table is illustrated in Figure 5.12. We developed sample software programs that target each instruction in the MicroBlaze processor instruction set by embedding assembly code into the sample C programs. In the embedded assembly code, we repeatedly execute the instruction of interest for a certain amount of time with more than 100 different sets of input data and under various execution contexts. ModelSim was used to perform low-level simulation for executing the sample software programs. The gate-level switching activities of the device during the execution of the sample software programs are recorded by ModelSim as simulation record files (.vcd files). Finally, low-level energy estimator such as XPower was used to analyze these simulation record files and estimate energy dissipation of the instructions of interest. See [71] for more details on the construction of instruction-level energy estimators for FPGA configured soft processors.

5.4.2.2   Domain-Specific Modeling Based Energy Estimation for Hardware Execution

Energy dissipation of the customized hardware peripherals is estimated through the integration of a domain-specific energy performance modeling technique proposed by Choi et al. [17]. As is shown in Figure 5.2, a *kernel* (a specific functionality performed by the customized hardware peripherals) can be implemented on FPGAs using different architectures and algorithms. For example, matrix multiplication on FPGAs can employ a single processor or a systolic architecture. FFT on FPGAs can adopt a radix-2 based or a radix-4

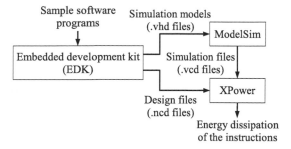

FIGURE 5.12: Flow for generating the instruction energy look-up table

based algorithm. These different architectures and algorithms use different amounts of logic components and interconnect, which prevent modeling their energy performance through a single high-level model.

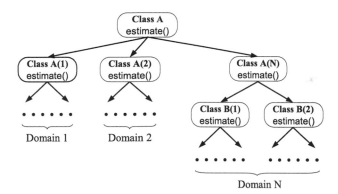

FIGURE 5.13: Python classes organized as domains

In order to support the domain-specific energy performance modeling technique, the application designer must be able to group different designs of the kernels into domains and associate the performance models identified through domain-specific modeling with the domains. Since the organization of the MATLAB/Simulink block set is inflexible and further re-organization and extension is difficult, we map the blocks in the Simulink block set into classes in the object-oriented Python scripting language [29]. As shown in Figure 5.14, the mapping follows some naming rules. For example, block *xbsBasic_r3/Mux*, which represents hardware multiplexers, is mapped to a Python class `CxlMul`. All the design parameters of this block, such as *inputs* (number of inputs), *precision* (precision), are mapped to the data attributes of its corresponding class and are accessible as `CxlMul.inputs` and `CxlMul.precision`. Information on the input and output ports of the blocks is stored in data attributes *ips* and *ops*. By doing so, hardware implementations are described using Python language and are automatically translated into corresponding designs in MAT-LAB/Simulink. For example, for two Python objects A and B, $A.ips[0:2] = B.ops[2:4]$ has the same effect as connecting the third and fourth output ports of the Simulink block represented by B to the first two input ports of the Simulink block represented by A.

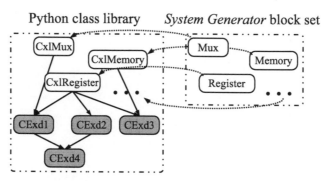

FIGURE 5.14:　Python class library

After mapping the block set to the flexible class library in Python, re-organization of the class hierarchy according to the architectures and algo-rithms represented by the classes becomes possible. Considering the example shown in Figure 5.13, Python class A represents various implementations of a kernel. It contains a number of subclasses $A(1)$, $A(2)$, $\cdots$, $A(N)$. Each of the subclasses represents one implementation of the kernel that belongs to the same *domain*. Energy performance models identified through domain-specific modeling (i.e. energy functions shown in Figure 5.2) are associated with these classes. Input to these energy functions is determined by the attributes of Python classes when they are instantiated. When invoked, the `estimate()` method associated with the Python classes returns the energy dissipation of the Simulink blocks calculated using the energy functions.

As a key factor that affects energy dissipation, switching activity informa-tion is required before these energy functions can accurately estimate energy dissipation of a design. The switching activity of the low-level implementations is estimated using the information obtained from the high-level co-simulation described in Section 5.4.1. For example, the switching activity of the Simulink block for addition is estimated as the average switching activity of the two input and the output data. The switching activity of the processing elements (PEs) of the CORDIC design shown in Figure 5.33 is calculated as the av-erage switching activity of all the wires that connect the Simulink blocks contained by the PEs. As shown in Figure 5.15, high-level switching activi-ties of the processing elements (PEs) shown in Figure 5.33 obtained within

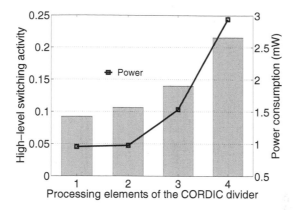

FIGURE 5.15:   High-level switching activities and power consumption of
the PEs shown in Figure 5.33

MATLAB/Simulink coincide with their power consumption obtained through
low-level simulation. Therefore, using such high-level switching activity estimates can greatly improve the accuracy of our energy estimates. Note that
for some Simulink blocks, the high-level switching activities may not coincide
with their power consumption under some circumstances. For example, Figure 5.16 illustrates the power consumption of slice-based multipliers for input
data sets with different switching activities. These multipliers demonstrate
"ceiling effects" when switching activities of the input data are larger than
0.23. Such "ceiling effects" are captured when deriving energy functions for
these Simulink blocks in order to ensure the accuracy of our rapid energy
estimates.

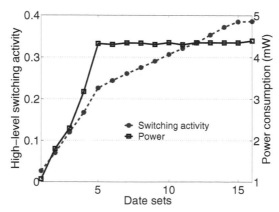

FIGURE 5.16:   High-level switching activities and power consumption of
slice-based multipliers

5.5 Energy Estimation for Customized Hardware Components

5.5.1 Software Architecture

We choose Python to implement the programmable interface *PyGen*. This is motivated by the fact that Python is an object-oriented scripting language with concise syntax, flexible data types, and dynamic typing [29]. It is widely used in many software systems. There are also attempts to use Python for hardware designs [38].

The software architecture of *PyGen* is shown in Figure 5.17. It contains four major modules. The architecture and the function of these modules are described in the following subsections.

5.5.1.1 *PyGen* Module

The *PyGen* module is a basic Python module, which is responsible for creating communication between *PyGen* and MATLAB/Simulink. It maps the basic building blocks in *System Generator* to Python classes, allowing users to create high-level hardware designs using the Python scripting language.

MATLAB provides three ways for creating such communication: MATLAB COM (Component Object Model) server, MATLAB engine, and a Java interface [60]. The communication interface is built through the MATLAB COM server by using the Python Win32 extensions from [40]. Through this interface, *PyGen* and *System Generator* can obtain the relevant information from each other and control the behavior of each other. For example, moving a design block in *System Generator* can change the placement properties of the corresponding Python object and vice versa. After a design is described

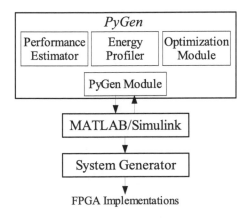

FIGURE 5.17: Architecture of *PyGen*

in Python, the end user can request the *PyGen* module to instruct MAT-LAB/Simulink through the COM interface and create a corresponding design in the Simulink modeling environment. The *PyGen* module is a fundamental module of *PyGen*. Application designers are required to import this module at the beginning using the Python command `import PyGen` whenever they need to instantiate their designs in Python.

Using some specific naming convention, the *PyGen* module maps the basic block set provided by *System Generator* to the corresponding classes (*basic classes*) in Python, which is shown in Figure 5.18. For example, block *xbsBasic_r3/Mux*, which is a *System Generator* block representing hardware multiplexers, is mapped to a Python class `CxlMul`. All the design parameters of this block, such as *inputs* (number of inputs) and *precision* (precision), are mapped to the data attributes of its corresponding class and are accessible as `CxlMul.inputs` and `CxlMul.precision`. The information on the input and output ports of the blocks is stored in data attribute *ips* and *ops*. Therefore, for two Python objects A and B, *A.ips[0:2] = B.ops[2:4]* has the same effect as connecting the third and fourth output ports of block B to the first two input ports of A.

Using the *PyGen* module, application designers describe their designs by instantiating classes from the Python class library, which is equivalent to dragging and dropping blocks from the *System Generator* block set to their designs. By leveraging the object-oriented class inheritance in Python, application designers can extend the class library by creating their own classes (*extended classes*, represented by the shaded blocks in Figure 5.18) and derive parameterized designs. This is further discussed in Section 5.5.3.1.

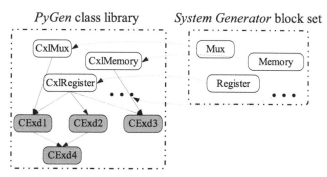

FIGURE 5.18:   Python class library within *PyGen*

5.5.1.2   Performance Estimator

Upon the instantiation of a *PyGen* class, a performance model is automatically generated from the performance model database and associated with the generated object. The performance model captures the performance of this object under the specific settings during instantiation, such as resource utilization, latency, and energy dissipation. The resource utilization is obtained by invoking the *Resource Estimator* block provided by *System Generator* and parsing its output. Currently, the hardware resources captured by *PyGen* include the number of slices, the amount of Block RAMs, and the number of embedded multipliers used in a design. The data processing latency of a *PyGen* is calculated based on two scenarios: if an object is instantiated directly from the basic class library, it is the *latency* data attribute of the object when it is instantiated from the basic classes; if an object is instantiated from the user extended class library, the performance estimator will traverse the object graph contained by the object and output the latency values based on the connectivity of the graph and the *latency* attribute of the objects. To obtain the energy performance, we integrate the domain-specific modeling technique described in Section 5.3 for rapid and accurate energy estimation. The energy estimation process is explained with more details in Section 5.5.3.2.

5.5.1.3   Energy Profiler

The energy profiler can analyze the low-level energy estimation results and reports the energy dissipation of a given *PyGen* object in a high-level design. The design flow using the profiler is shown in Figure 5.19. The energy profiling approach used by the energy profiler is similar to the timing analyzer provided by *System Generator*. After the design is created, the application designers follow the standard FPGA design flow to synthesize and implement the design. Design files (.ncd files) that represent the FPGA netlists are generated. Then, it is simulated using ModelSim to generate simulation files (.vcd files). These files record the switching activity of the various hardware components on the device. The design files (.vcd files) and the simulation files (.vcd files) are then fed back to the profiler within *PyGen*. The profiler has an interface with XPower [97] and can obtain the average power consumption of the clock

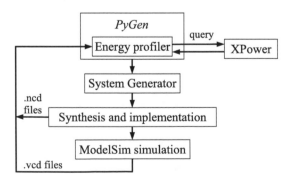

FIGURE 5.19:   Design flow using the energy profiler

network, nets, logic, and I/O pads by querying XPower through this interface. The energy profiler instructs *System Generator* to maintain the naming hierarchy of the generated HDL designs and netlists. Using the retained naming hierarchy, the energy profiler can identify the low-level hardware components and wires that belong to a specific *System Generator* block. By summing up these power values of these hardware components and wires, the energy profiler outputs the power consumption of *System Generator* blocks or *PyGen* objects. Combining with appropriate timing information, the power values can be further translated to values of energy dissipation.

The energy profiler can help to identify the energy hot spots in designs. More importantly, as discussed in Section 5.1, it can be used to generate feedback information for improving the accuracy of the performance estimator.

5.5.1.4 Optimization Module

The optimization module is a programmable interface, which allows the application designers to access the domain-specific models and performance data and implement various algorithms to identify the optimized designs. This module offers two basic functionalities: description of the design constraints and optimization of the design with respect to the constraints. Since parameterized designs are developed as Python classes, application designers realize the two functions by writing Python code and manipulating the *PyGen* classes. This gives the designers complete flexibility to incorporate a variety of optimization algorithms and provides a way to quickly traverse the MATLAB/Simulink design space.

5.5.2 Overall Design Flow

Based on the architecture of *PyGen* discussed above, the design flow is illustrated in Figure 5.20. The shaded boxes represent the four major functionalities offered by *PyGen* in addition to the original MATLAB/Simulink design flow.

- *Parameterized design development.* Parameterized designs are described in Python. Design parameters, such as data precision, degree of parallelism, hardware binding, etc., can be captured by the Python designs. After the designs are completed, *PyGen* is invoked to translate the designs in Python to the corresponding designs in MATLAB/Simulink. Changes to the MATLAB/Simulink designs, such as the adjustment of the placement of the blocks, also get reflected in the *PyGen* environment through the communication channel between them.

- *Performance estimation.* Using the modeling environment of MATLAB/Simulink, application designers can perform arithmetic level simulation to verify the correctness of their designs. Then, by providing the simulation results to the performance estimator within *PyGen* and invoking it, application designers can quickly estimate the performance of their designs,

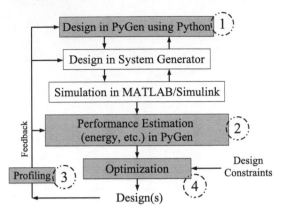

FIGURE 5.20:   Design flow of *PyGen*

such as energy dissipation and resource utilization.

• *Optimization for energy efficiency.* Application designers provide design constraints, such as end-to-end latency, throughput, number of available slices and embedded multipliers, etc., to the optimization module. After optimization is completed, *PyGen* outputs the designs which have the maximum energy efficiency according to the performance metrics used while satisfying the design requirements.

• *Profile and feedback.* The design process can be iterative. Using the energy profiler, *PyGen* can break down the results from low-level simulation and profile energy dissipation of various components of the candidate designs. The application designers can use this profiling to adjust the architectures and algorithms used in their designs. Such energy profiling information can also be used to refine the energy estimates from the performance estimator.

Finally, using *System Generator* to generate the corresponding VHDL code, application designers can follow the standard FPGA design flow to synthesize and implement these designs and download them to the target devices.

The input to our design tool is a task graph. That is, the target application is decomposed into a set of tasks with communication between them. Then, the development using *PyGen* is divided into two levels: kernel level and application level. The objectives of kernel level development are to develop parameterized designs for each task and to provide support for rapid energy estimation. The objectives of application level development are to describe the application using the available kernels and to optimize its energy performance with respect to design constraints.

5.5.3   Kernel Level Development

The kernel level development consists of two design steps, which are discussed below.

5.5.3.1 Parametrized Kernel Development

As shown in [17], different implementations of a task (e.g., kernel) provides different design trade-offs for application development. Taking matrix multiplication as an example, designs with a lower degree of parallelism require less hardware resources than those with a higher degree of parallelism while introducing a larger latency. Also, at the implementation level, several trade-offs are available. For example, in the realization of storage, registers, slice-based RAMs and Block RAMs can be used. These implementations offer different energy efficiency depending on the size of data that needs to be stored. The objective of parameterized kernel design is to capture these design and implementation trade-offs and make them available for application development.

While *System Generator* offers limited support for developing parameterized kernels, *PyGen* has a systematic mechanism for this purpose by the way of Python classes. Application designers expand the Python class library and create *extended classes*. Each *extended class* is constructed as a tree, which contains a hierarchy of subclasses. The leaf nodes of the tree are *basic classes* while the other nodes are *extended classes*. An example of such an *extended class* is shown in Figure 5.21. This example illustrates some extended classes in the construction of a parameterized FFT kernel in *PyGen*. Once an *extended class* is instantiated, its subclasses also get instantiated. Translating this to the MATLAB/Simulink environment by the *PyGen* module has the same effect as generating *subsystems* in MATLAB/Simulink, dragging and dropping a number of blocks into these *subsystems*, and connecting the blocks and the *subsystems* according to the relationship between the classes.

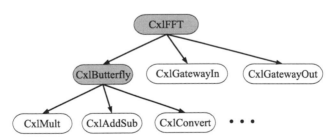

FIGURE 5.21: Tree structure of the Python extended classes for parameterized FFT kernel development

Application designers are interested in some design parameters while generating the kernels. These parameters can be architecture used, hardware binding of a specific function, data precision, degree of parallelism, etc. We use the data attributes of the Python classes to capture these design parameters. Each design parameter of interest has a corresponding data attribute in the Python class. These data attributes control the behavior of the Python class when the class is instantiated to generate *System Generator* designs. They determine the blocks used in a MATLAB/Simulink design and the connections between the blocks. Besides, by properly packaging the classes, the application designers can choose to expose only the data attributes of interest for application level development.

5.5.3.2   Support of Rapid and Accurate Energy Estimation

While the parameterized kernel development can potentially offer a large design space, being able to quickly and accurately obtain the performance of a given kernel is crucial for identifying the appropriate parameters of the kernel and optimize the performance of the application using it. To address this issue, we integrate into *PyGen* a domain-specific modeling based rapid energy estimation technique proposed in [17].

The use of domain-specific energy modeling for FPGAs is discussed in details in Section 5.3. In general, a kernel can be implemented using different architectures. For example, implementing matrix multiplication on FPGAs can employ a single processor or a systolic architecture. Implementations using a particular architecture are grouped into a domain. Analysis of energy dissipation of the kernel is performed within each domain. Because each domain corresponds to an architecture, energy functions can be derived for each domain. These functions are used for rapid energy estimation for implementations in the corresponding domain. See [17] for more details regarding domain-specific modeling.

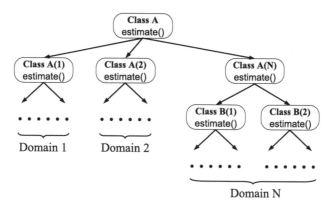

FIGURE 5.22:   Class tree organized as domains

In order to support this domain-specific modeling technique, the kernel developers must be able to group different kernel designs into the corresponding domain. Such support is not available in *System Generator* as the organization of the block set is fixed. However, after mapping the block set to the flexible class library in *PyGen*, re-organization of the class hierarchy according to the architectures represented by the classes becomes possible. Taking the case shown in Figure 5.22 as an example, Python class A represents various implementations of a kernel. It contains a number of subclasses $A(1)$, $A(2)$, $\cdots$, $A(N)$. Each of the subclasses represents the implementations of the kernel that belong to the same *domain*.

The process of energy estimation in *PyGen* is hierarchical. Energy functions are associated with the Python *basic classes* and are obtained through low-level simulation. They capture the energy performance of these basic classes under various possible parameter settings. For the *extended classes*, depending on whether or not domain-specific energy modeling is performed for the classes, there may be no energy functions associated with them for energy estimation. In case that the energy function is not available, energy estimate of the class needs to be obtained from the classes contained in it. While this way of estimation is fast because it skips the derivation of energy functions, it has lower estimation accuracy as shown in Table 5.6 in Section 5.7.

To support such a hierarchical estimation process, a method `estimate()` is associated with each Python object. When this method is invoked, it checks if an energy function is associated with the Python object. If yes, it calculates the energy dissipation of this object according to the energy function and the parameter settings for this object. Otherwise, *PyGen* iteratively searches the tree as shown in Figure 5.21 within this Python object until enough information is obtained to calculate the energy performance of the object. In the worst case, it will trace all the way back to the leaf nodes of the tree. Then, the `estimate()` method computes the energy performance of the Python object using the energy functions obtained as described above.

Switching activities within a design is a key factor that affects energy dissipation. By utilizing the data from MATLAB/Simulink simulation, *PyGen* obtains the *actual* switching activity of various blocks in the high level designs and uses them for energy estimation. Compared to the approach in [17] which assumes default switching activities, this approach helps increase the accuracy of the estimates. To show the benefits offered by *PyGen*, we consider an 8-point FFT using the unfolded architecture discussed in Section 5.7. It contains twelve butterflies, each based on the same architecture. In Figure 5.23, the bars show the power consumption of these butterflies while the upper curve shows the average switching activities of the *System Generator* basic building blocks used by each butterfly. Such switching activity information can be *quickly* obtained from the MATLAB/Simulink arithmetic level simulation. As shown in Figure 5.23, the switching activity information obtained from MATLAB/Simulink is able to capture the variation of the power consumption of these butterflies. The average estimation error based on such switching activity information is 2.9%. For the sake of comparison, we perform energy estimation by assuming a default switching activity as in [17]. The results are shown in Figure 5.24. For default switching activities ranging from 20% to 40%, which are typical of designs of many signal processing applications, the average estimation errors can go up to as much as 36.5%. Thus, by utilizing the MATLAB/Simulink simulation results, *PyGen* improves the estimation accuracy.

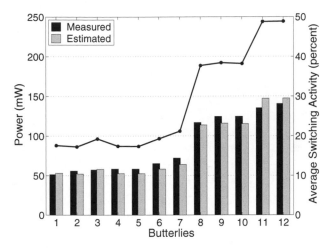

FIGURE 5.23:   Power consumption and average switching activities of input/output data of the butterflies in an unfolded-architecture for 8-point FFT computation

5.5.4   Application Level Development

The application level development begins after the parameterized designs for the tasks are made available by going through the kernel level development. It consists of three design steps, which are discussed below.

5.5.4.1   Application Description

Based on the input task graph, the application designers construct the application using the parameterized kernels as discussed in the previous section. This is accomplished by manipulating the Python classes created in a way as described in Section 5.5.3.1. Application designers also need to create interfacing classes for describing the communication between tasks. These classes capture: (1) data buffering requirement between the tasks, which is determined by the application requirements and the data transmission patterns of the implementations of the tasks; (2) hardware binding of the buffering.

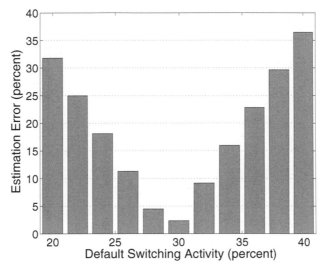

FIGURE 5.24:   Estimation error of the butterflies when default switching activity is used

5.5.4.2   Support of Performance Optimization

Application designers have complete flexibility in implementing the optimization module by handling the Python object. For example, if the task graph of the application is a linear pipeline, the application designer can create a trellis as shown in Figure 5.25. Many signal processing applications including the beamforming application discussed in the next section can be described as linear pipelines. The parameterized kernel classes capture the various possible implementations of the tasks (the shaded circles on the trellis) while the interfacing classes capture the various possible ways of communication between the tasks (the connection between the shaded circles on the trellis). Then, the dynamic programing algorithm proposed in [68] can be applied to find out the design parameters for the tasks so that the energy dissipation of executing one data sample is minimized.

5.5.4.3   Energy Profiling

By using the energy profiler in *PyGen*, application designers can write Python code to obtain the power or energy dissipation for a specific Python object or a specific kind of object. For example, the power consumption of the butterflies used in an FFT design is shown in Figure 5.23. Based on the profiling, the application designers can identify the energy *hot spots* and change the designs of the kernels or the task graph of the applications to further increase energy efficiency of their designs. They can also use the profiling to refine the energy estimates from the energy estimator. One major reason that

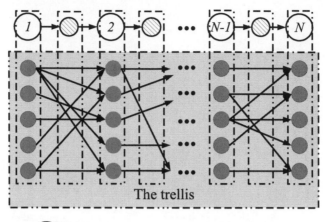

(1)   Kernel classes implementing the tasks

(⊘)   Interfacing classes capturing the
        communication between tasks

FIGURE 5.25:   Trellis for describing linear pipeline applications

necessitates such refinement is that the energy estimation using the energy
functions (discussed in Section 5.5.3.2) captures the energy dissipation of the
Python objects; it cannot capture the energy dissipated by the interconnect
that provides communication between these objects.

5.6   Instruction-Level Energy Estimation for Software Programs

5.6.1   Arithmetic-Level Instruction Based Energy Estimation

In this chapter, we propose an arithmetic-level instruction based energy
estimation technique which can rapidly and fairly accurately estimate the
energy dissipation of computations on FPGA based soft processors. We use
instruction based to denote that the assembly instructions that constitute the
input software programs are the basic units based on which the proposed
technique performs energy estimation. The energy dissipation of the input
software program is obtained as the sum of the energy dissipation of the
assembly instructions executed when the software program is running on the
soft processor. The architecture of many soft processors consists of multiple
pipeline stages in order to improve their time performance. For example, the

MicroBlaze processor contains three pipeline stages. While most instructions take one clock cycle to complete, the memory access *load/store* instructions take two clock cycles to complete and the multiplication instruction takes three clock cycles to complete. Thus, the execution of instructions in these soft processors may get stalled due to the unavailability of the required data. We assume that the various hardware components composing the soft processor are "properly" clock gated between the executions of the assembly instructions (e.g., the MicroBlaze processor) and dissipate negligible amounts of energy when they are not in use. Under this assumption, the stalling of execution in the soft processor would have negligible effect on the energy dissipation of the instructions. Thus, the impact of multiple pipeline stages on the energy dissipation of the instructions is ignored in our energy estimation technique.

We use *arithmetic-level* to denote that we capture the changes in the arithmetic status of the processor (i.e., the arithmetic contents of the registers and the local memory space) as a result of executing the instructions. We use this *arithmetic-level* information to improve the accuracy of our energy estimates. We do not capture how the instructions are actually carried out in the functional units of the soft processor. Instead, the impact of different functional units on the energy dissipation is captured indirectly when analyzing the energy performance of the instructions. For example, when analyzing the energy dissipation of the addition instructions, we compare the values of the destination register before and after the addition operation and calculate the switching activity of the destination register as a result of the addition operation. Sample software programs which cause different switching activities on the destination register are executed on the soft processor. The energy dissipation of these sample software programs is obtained through low-level energy estimation and/or actual measurement. Then, the energy performance of the addition instructions is expressed as an *energy function* of the switching activity of the destination register of the addition instruction. As we show in Section 5.6.3, this arithmetic-level information can be used to improve the accuracy when estimating the energy dissipation of the software programs. One major advantage of such arithmetic-level energy estimation is that it avoids the time-consuming low-level simulation process and greatly speeds up the energy estimation process. Note that since most soft processors do not support out-of-order execution, we only consider the change in arithmetic status of the processor between two immediately executed instructions when analyzing the energy dissipation of the instructions.

Based on the above discussion, our rapid energy estimation technique for soft processors consists of two steps. In the first step, by analyzing the energy performance of the sample programs that represent the corresponding instructions, we derive energy functions for all the instructions in the instruction set of the soft processor. An arithmetic-level instruction energy look-up table is created by integrating the energy functions for the instructions. The look-up table contains the energy dissipation of the instructions of the soft processor executed under various arithmetic status of the soft processor.

In the second step, an instruction set simulator is used to obtain the information of the instructions executed during the execution of the input software program. This information includes the order that the instructions are executed and the arithmetic status of the soft processor (e.g., the values stored at the registers) during the execution of the instructions. For each instruction executed, its impact on the arithmetic status of the soft processor is obtained by comparing the arithmetic contents of the registers and/or the local memory space of the soft processor before and after executing the instruction. Then, based on this arithmetic-level information, the energy dissipation of the executed instructions is obtained using the instruction energy look-up table. Finally, the energy dissipation of the input software program is obtained by summing up the total energy dissipation of all the instructions executed in the software program.

5.6.2   An Implementation

An implementation of the proposed arithmetic-level instruction set based energy estimation technique, which is shown in Figure 5.27, is provided to illustrate our energy estimation technique. While our technique can be applied to other soft processors, Xilinx MicroBlaze is used to demonstrate the energy estimation process since this processor has been widely used.

5.6.2.1   Step 1 — Creation of the Arithmetic-Level Instruction Energy Look-Up Table

The flow for creating the instruction energy look-up table is shown in Figure 5.26. We created sample software programs that represent each instruction in the MicroBlaze instruction set by embedding assembly code into the sample C programs. For the embedded assembly code, we let the instruction of interest be repeatedly executed for a certain amount of time with more than 100 different sets of input data and under various execution contexts (e.g., various orders of instruction execution orders). More specifically, we use *mb-gcc*, the GNU *gcc* compiler for MicroBlaze for compiling the input software program. Then, *mb-gdb* (the GNU *gdb* debugger for MicroBlaze), the Xilinx Microprocessor Debugger (XMD), and the MicroBlaze cycle-accurate instruction set simulator provided by Xilinx EDK (Embedded Development Kit) [97] are used to obtain the arithmetic-level instruction execution information of the software program. Following the design flows discussed in Section 5.6.1, the instruction energy look-up table that stores the energy dissipation of each instruction in the instruction set of the MicroBlaze processor is created.

We use low-level simulation based technique to analyze the energy performance of the instructions. The MicroBlaze processor has a *PC_EX* bus which displays the value of the program counter and a *VALID_INSTR* signal which is high when the value on the *PC_EX* bus is valid. We first perform timing based gate-level simulation of the sample software programs using ModelSim [63].

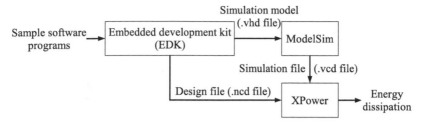

FIGURE 5.26:   Flow of instruction energy profiling

We check the status of *PC_EX* and *VALID_INSTR* during the timing based simulation. Using this information, we determine the intervals over which the instructions of interest are executed. Then, we rerun the timing based gate-level simulation in ModelSim. Simulation files (.vcd files) that record the switching activities of each logic and wire on the FPGA device during these intervals are generated. In the last step, XPower is used to analyze the design files (.ncd files) and the simulation files. Finally, the energy dissipation of the instructions is obtained based on the output from XPower.

Note that the MicroBlaze soft processor is shipped as a register-transfer level IP (Intellectual Property) core. Its time and energy performance are independent of its placement on the FPGA device. Moreover, the instructions are stored in the pre-compiled memory blocks (i.e., BRAMs). Since the BRAMs are evenly distributed over the FPGA devices, the physical locations of the memory blocks and the soft processor have negligible impact on the energy dissipation for the soft processor to access the instructions and data stored in the BRAMs. Therefore, as long as the MicroBlaze system is configured as in Figure 5.28, energy estimation using the instruction energy look-up table can lead to fairly accurate energy estimates, regardless of the actual placement of the system on the FPGA device. There is no need to incur the time-consuming low-level simulation each time the instruction energy look-up table is used for energy estimation in a new configuration of the hardware platform.

5.6.2.2   Step 2 — Creation of the Energy Estimator

The energy estimator created based on the two-step methodology discussed in Section 5.6.1 is shown in Figure 5.27. It is created by integrating several software tools. Our main contributions are the arithmetic-level instruction energy look-up table and the instruction energy profiler, which are highlighted as shaded boxes in Figure 5.27. The creation of the instruction energy is discussed in Section 5.6.2.1. For the instruction energy profiler, it integrates the instruction energy look-up table based on the execution of the instructions and calculates the energy dissipation of the input software program. Note that, as a requirement of the MicroBlaze cycle-accurate instruction set simulator, the configurations of the soft processor, its memory controller interfaces,

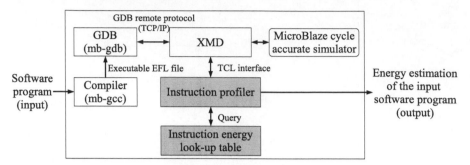

FIGURE 5.27:   Software architecture of an implementation of the
proposed energy estimation technique

and memory blocks (e.g., BRAMs) are pre-determined and are shown in Figure 5.28. Since MicroBlaze supports split bus transactions, the instructions and the data of the user software programs are stored in the dual port BRAMs and are made available to the MicroBlaze processor through two separate LMB (Local Memory Bus) interface controllers. The MicroBlaze processor and the two LMB interface controllers are required to operate at the same frequency. Under this setting, a fixed latency of one clock cycle is guaranteed for the soft processor to access the program instructions and data through these two memory interface controllers.

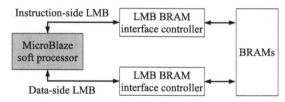

FIGURE 5.28:   Configuration of the MicroBlaze processor system

The GNU compiler for MicroBlaze, *mb-gcc*, is used to compile the software program and generate an ELF (Executable and Linking Format) file, which can be downloaded to and executed on the MicroBlaze processor. This ELF file is then provided to the GNU debugger for MicroBlaze, *mb-gdb*.

The *Xilinx Microprocessor Debugger* (XMD) is a tool from Xilinx [97] for debugging programs and verifying systems using the PowerPC (Virtex-II Pro) or MicroBlaze microprocessors. It contains a GDB interface. Through this interface, XMD can communicate with *mb-gdb* using TCP/IP protocol and get access to the executable ELF file that resides in it. XMD also integrates a built-in cycle accurate simulator that can simulate the execution of instructions within the MicroBlaze processor. The simulator assumes that all the instructions and program data are fetched through two separate local memory bus (LMB) interface controllers. Currently, the simulation of the MicroBlaze processor with other configurations, such as different bus protocols and other peripherals, is not supported by this instruction set simulator.

XMD has a TCL (Tool Command Language) [26] scripting interface. By communicating with XMD through the TCL interface, the instruction profiler obtains the numbers and the types of instructions executed on the MicroBlaze processor. Then, the profiler queries the instruction energy look-up table to find out the energy dissipation of each instruction executed. By summing up all the energy values, the energy dissipation of the input software program is obtained.

5.6.3   Illustrative Examples

The MicroBlaze based system shown in Figure 5.28 is configured on a Xilinx Spartan-3 XC3S400 FPGA device, which integrates dedicated 18bit×18bit multipliers and BRAMs. We configured MicroBlaze to use three dedicated multipliers for the multiplication instructions (e.g., *mul* and *muli*) and to use the BRAMs to store the instructions and data of the software programs. According to the requirement of the cycle-accurate simulator, the operating frequencies of the MicroBlaze processor and the two LMB interface controllers are set at 50 MHz.

For the experiments discussed in the chapter, we use EDK 6.2.03 for description and automatic generation of the hardware platforms. The GNU tool chain for MicroBlaze (e.g., *mb-gcc* and *mb-gdb*) from Xilinx [97] is used for compiling the software programs. We also use the Xilinx ISE (Integrated Software Environments) tool 6.2.03 [97] for synthesis and implementation of the hardware platform and ModelSim 6.0 [63] for low-level simulation and recording of the simulation results. The functional correctness and the actual power consumption of the designs considered in our experiments are verified on a Spartan-3 FPGA prototyping board from Nu Horizons [66].

5.6.3.1   Creation of the Arithmetic-Level Instruction Energy Look-Up Table

Using the technique discussed in Section 5.6.2.1, we perform low-level energy estimation to create the arithmetic-level instruction energy look-up table for the MicroBlaze processor.

• *Variances in energy dissipation of the instructions*

Figure 5.29 shows the energy profiling of various MicroBlaze instructions obtained using the methodology described in Section 5.6.2.1. We consider the energy dissipated by the complete system shown in Figure 5.28, which includes the MicroBlaze processor, two memory controllers, and the BRAMs. Instructions for addition and subtraction have the highest energy dissipation while memory access instructions (both load and store) and branch instructions

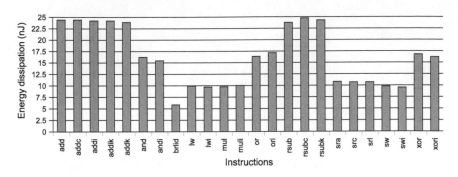

FIGURE 5.29:   Energy profiling of the MicroBlaze instruction set

have the lowest energy dissipation. A variance of 423% in energy dissipation is observed for the entire instruction set, which is much higher than the 38% variance for both StrongARM and Hitachi SH-4 processors reported in [85]. These data justify our motivation for using an approach based on instruction-level energy profiling to build the energy estimator for soft processors, rather than using the first order and the second order models proposed in [85].

Note that the addition and subtraction instructions (e.g., *add, addi, sub,* and *rsubik*) dissipate much more energy than the multiplication instructions. This is because in our configuration of the soft processor, addition and subtraction are implemented using "pure" FPGA resources (i.e., slices on Xilinx FPGAs) while multiplication instructions are realized using the embedded multipliers available on the Spartan-3 FPGAs. Our work in [17] shows that the use of these embedded multipliers can substantially reduce the energy dissipation of the designs compared with those implemented using configurable logic components.

• *Impact of input data causing different arithmetic behavior of the soft processor*

Figure 5.30 shows the energy dissipation of the *add* addition instruction, the *xor* instruction, and the *mul* multiplication instruction. They are executed with different input data which cause different switching activities on the destination registers. We can see that input data causing different switching activities of the soft processor have a significant impact on the energy dissipation of these instructions. Thus, by utilizing the arithmetic-level information gathered during the instruction set simulation process, our rapid energy estimation technique can greatly improve the accuracy of energy estimation.

• *Impact of various addressing modes*

We have also analyzed the impact of various addressing modes of the instructions on their energy dissipation. For example, consider adding the content of register *r1* and number 23 and storing the result in register *r3*. This can be accomplished using the immediate addressing mode of the addition

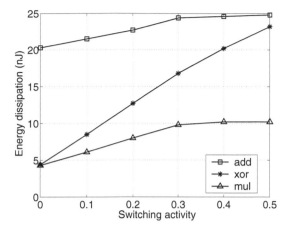

FIGURE 5.30:   Impact of input data causing different arithmetic behavior of MicroBlaze

instruction as *addi r3, r1, 23* or using the indirect addressing mode of the addition instruction as *add r3, r1, r16* with the content of register *r16* being 16 before executing the addition instruction. The energy dissipation of the addition instruction in different addressing modes is shown in Figure 5.31. Compared with instruction decoding and the actual computation, different addressing modes contribute little to the overall energy dissipation of the instructions. Thus, different addressing modes of the instructions can be ignored in the energy estimation process in order to reduce the time for energy estimation.

5.6.3.2   Energy Estimation of the Sample Software Programs

• *Sample Software Programs*

In order to demonstrate the effectiveness of our approach, we analyze the energy performance of two FFT software programs and two matrix multiplication software programs running on the MicroBlaze processor system configured as shown in Figure 5.28. Details of these four software programs can be found in [77].

We choose FFT and matrix multiplication as our illustrative examples because they are widely used in many embedded signal processing systems, such as software defined radio [64]. For floating-point computation, there are efforts in supporting basic operations on FPGAs, such as addition/subtraction, multiplication, division, etc. [36] [23]. These implementations require a significant amount of FPGA resources and involve considerable effort to realize floating-point FFT and matrix multiplication on FPGAs. In contrast, floating-point FFT and matrix multiplication can be easily implemented on soft processors through software emulation, requiring only a moderate amount of memory space.

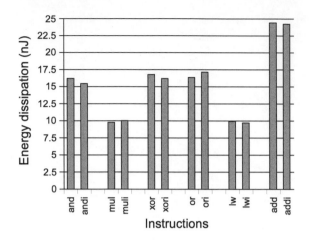

FIGURE 5.31:   Impact of different instruction addressing modes

- *The MicroBlaze soft processor platform*

By time sharing the resources, implementing FFT and matrix multiplication using soft processors usually demands a smaller amount of resources than the corresponding designs with parallel architectures. Thus, they can be fit into smaller devices. The complete MicroBlaze system occupies 736 (21%) of 3584 available slices, 3 (18.8%) of 16 available dedicated multipliers on the target Spartan-3 FPGA device. The floating-point FFT program occupies 23992 bits while the floating-point matrix multiplication program occupies 3976 bits. Each of the four software programs can be fit into 2 (12.5%) of 16 available 18-Kbit BRAMs. While parallel architectures reduce the latency for floating-point matrix multiplication, they require 33 to 58 times more slices (19045 and 33589 slices) than the MicroBlaze based software implementation [119]. These customized designs with parallel architecture cannot fit in the small Spartan-3 FPGA device targeted in this chapter. As quiescent power accounts for increasingly more percentage of the overall power consumption on modern FPGAs [93], choosing a smaller device can effectively reduce the quiescent energy. For example, according the Xilinx web power tools [103], the quiescent power of the Spartan-3 target device used in our experiments is 92 mW. It is significantly smaller than the 545 mW quiescent power of the Virtex-II Pro XC2VP70 device targeted by Zhuo and Prasanna's designs [119].

- *Energy performance*

The energy dissipation of the FFT programs and the matrix multiplication programs estimated using our technique is shown in Table 5.5. All these estimates were obtained within five minutes. Our arithmetic-level instruction based energy estimation technique eliminates the ~6 hours required for register-transfer level low-level simulation of the soft processor and the ad-

TABLE 5.5: Energy dissipation of the FFT and matrix multiplication software programs

Program	Measured	Low-level estimation	Our technique
FFT	1783 J	1662 J (6.8%)	1544 J (13.4%)
	300 J	273 J (3.2%)	264 J (12.0%)
Matrix	180 J	162 J (8.5%)	153 J (15.0%)
Multiplication	170 J	157 J (7.8%)	145 J (14.7%)

Program	No. of clock cycles	Comment
FFT	97766	8-point, complex floating-point data
	16447	8-point, complex integer data
Matrix	9877	3×3 matrix, real floating-point data
Multiplication	9344	3×3 matrix, real integer data

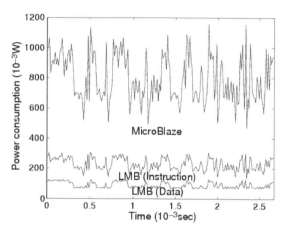

FIGURE 5.32: Instant average power consumption of the FFT software program

ditional ~3 hours required for analyzing the low-level simulation results to obtain the energy estimation results as discussed in Section 5.1. Thus, our technique achieves a significant speed-up compared with the low-level simulation based energy estimation techniques.

We have performed low-level simulation based energy estimation using XPower to estimate the energy dissipation of these software programs. Also, to verify the energy dissipation results obtained using our arithmetic-level rapid estimation technique, we have also performed actual power consumption measurement of the MicroBlaze soft processor platform using a Spartan-3 prototyping board from Nu Horizons [66] and a SourceMeter 2400 from Keithley [53]. During our measurement, we ensure that except for the Spartan-3 FPGA chip, all the other components on the prototyping board (e.g., the power supply indicator and the SRAM chip) are kept in the same operating state when MicroBlaze processor is executing the software programs. Under these settings, we consider that the changes in power consumption of the FPGA prototyping board are mainly caused by the FPGA chip. Using the Keithley SourceMeter, we fix the input voltage to the FPGA prototyping board at 6 Volts and measure the changes of input current to it. Dynamic power consumption of the MicroBlaze soft processor system is then calculated based on the changes of input current. Compared with measured data, our arithmetic-level energy estimation technique achieves estimation errors ranging from 12.0% to 15.0% and 13.8% on average. Compared with the data obtained through low-level simulation based energy estimation, our technique introduces an average estimation error of 7.2% while significantly speeding up the energy estimation process.

In addition, we have further analyzed the variation of the power consumption of the soft processor during the execution of the floating-point FFT program. First, we use ModelSim to generate a simulation record file every 13.3293 μsec (1/200 of the total simulation time). Then, XPower is used to measure the average power consumption of each period represented by these simulation record files. The results are shown in Figure 5.32. The regions between the lines represent the energy dissipation of the MicroBlaze processor, the instruction access cost, and the program data access cost (both through the LMB bus).

For the above floating-point FFT software program, the MicroBlaze processor dissipates around 3 times as much energy as that for accessing the instructions and data stored in the BRAMs. Most importantly, this result shows significant fluctuations in the power consumption of the entire system. This is consistent with the large variations in the energy profiling of the MicroBlaze instruction set shown in Figure 5.29. As different instructions are executed in each sampling period, the large differences in energy dissipation among the instructions would result in significant variation in the average power consumption in these sampling periods.

5.7   Illustrative Examples

To demonstrate the effectiveness of our approach, we show in this section the design of a CORDIC processor for division and a block matrix multiplication algorithm. These designs are widely used in systems such as software defined radio, where energy is an important performance metric [24]. We focus on MicroBlaze and *System Generator* in our illustrative examples due to their wide availability though our approach is also applicable to other soft processors and other design tools.

TABLE 5.6: Arithmetic level/low-level simulation time and measured/estimated energy performance of the CORDIC based division application and the block matrix multiplication application

Designs	Simulation time	
	Arithmetic level	Low-level*
CORDIC with $N = 24$, $P = 2$	6.3 sec	35.5 sec
CORDIC with $N = 24$, $P = 4$	3.1 sec	34.0 sec
CORDIC with $N = 24$, $P = 6$	2.2 sec	33.5 sec
CORDIC with $N = 24$, $P = 8$	1.7 sec	33.0 sec
12×12 matrix mult. (2×2 blocks)	99.4 sec	8803 sec
12×12 matrix mult. (4×4 blocks)	51.0 sec	3603 sec

Energy performance		
High-level	Low-level	Measured
1.15 μJ (9.7%)	1.19 μJ (6.8%)	1.28 μJ
0.69 μJ (9.5%)	0.71 μJ (6.8%)	0.76 μJ
0.55 μJ (10.1%)	0.57 μJ (7.0%)	0.61 μJ
0.48 μJ (9.8%)	0.50 μJ (6.5%)	0.53 μJ
595.9 μJ (18.2%)	675.3 μJ (7.3%)	728.5 μJ
327.5 μJ (12.2%)	349.5 μJ (6.3%)	373.0 μJ

*Note: Timing based post place-and-route simulation. The times for placing-and-routing and generating simulation models are not included.

• *CORDIC processor for division*: The CORDIC (COordinate Rotation DIgital Computer) iterative algorithm for dividing b by a [6] is described as follows. Initially, we set $X_{-1} = a$, $Y_{-1} = b$, $Z_{-1} = 0$ and $C_{-1} = 1$. There are N iterations and during each iteration i ($i = 0, 1, \cdots, N-1$), the following computation is performed.

$$\begin{cases} X_i = X_{i-1} \\ Y_i = Y_{i-1} + d_i \cdot X_{i-1} \cdot C_{i-1} \\ Z_i = Z_{i-1} - d_i \cdot C_{i-1} \\ C_i = C_{i-1} \cdot 2^{-1} \end{cases} \qquad (5.1)$$

where, $d_i = +1$ if $Y_i < 0$ and -1 otherwise. The customized hardware peripheral is described in MATLAB/Simulink as a linear pipeline of P processing elements (PEs). Each of the PEs performs one iteration of computation described in Equation 5.1. The software program controls the data flowing through the PEs and ensures that the data get processed repeatedly by them until the required number of iterations is completed. Communication between the processor and the hardware implementation is through the FSL interfaces. It is simulated using our MicroBlaze Simulink block. We also consider 32-bit data precision.

• *Block matrix multiplication*: Input matrices A and B are decomposed into a number of smaller matrix blocks. Multiplication of these smaller matrix

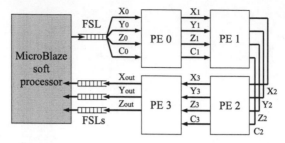

FIGURE 5.33:   CORDIC processor for division $(P = 4)$

TABLE 5.7:   Simulation speeds of various simulators (unit: number of clock cycles simulated per second)

Instruction set simulator	Simulink[1]	ModelSim[2]
>1000	254.0	8.7

Note: (1) Only consider simulation of the customized hardware peripherals; (2) Timing based post place-and-route simulation. The time for generating the simulation models of the low-level implementations is not accounted for.

blocks is performed using a customized hardware peripheral described using MATLAB/Simulink. As shown in Figure 5.34, data elements of a matrix block from matrix B (e.g. b_{11}, b_{21}, b_{12} and b_{22}) are fed into the hardware peripheral and get stored. When data elements of a matrix block from matrix A come in, multiplication and accumulation are performed accordingly to generate output results. The software program running on MicroBlaze is responsible for preparing the data sent to the customized hardware peripheral, accumulating the multiplication results back from the hardware peripheral, and generating the result matrix.

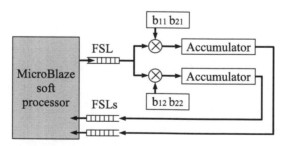

FIGURE 5.34:   Architecture of matrix multiplication with customized hardware for multiplying 2×2 matrix blocks

For the experiments discussed in this chapter, the MicroBlaze processor is configured on a Xilinx Spartan-3 xc3s400 FPGA [97]. The operating frequencies of the processor, the two LMB (Local Memory Bus) interface controllers, and the customized hardware peripherals shown in Figure 5.1 are set at 50 MHz. We use EDK 6.3.02 for describing the software execution platform and for compiling the software programs. *System Generator* 6.3 is used for description of the customized hardware peripherals and automatic generation of low-level implementations. Finally, ISE 6.3.02 [97] is used for synthesizing and implementing (placing and routing) the complete applications.

We have also measured the actual power consumption of the two applications using a Spartan-3 prototyping board from Nu Horizons [66] and a SourceMeter 2400 instrument (a programmable power source with the measurement functions of a digital multimeter) from Keithley [53]. Except for the Spartan-3 FPGA device, all the other components on the prototyping board (e.g., the power supply indicator and the SRAM chip) are kept in the same operating state when the FPGA device is executing the applications. Under these settings, we consider that the changes in power consumption of the board are mainly caused by the FPGA device. Then, we fix the input voltage and measure the changes of input current to the prototyping board. The dynamic power consumption of the designs is calculated based on the changes of input current. Note that *quiescent power* (power consumption of the device when there is no switching activity on it) is ignored in our experimental results since it can not be optimized for a specific FPGA device.

The simulation time and energy performance for various implementations of the two numerical computation applications are shown in Table 5.6. For these two applications, our arithmetic level co-simulation environment based on MATLAB/Simulink achieves simulation speed-ups ranging from 5.6x to 88.5x compared with that of low-level timing simulation using ModelSim. The low-level timing simulation is required for low-level energy estimation using XPower. The simulation speed of our arithmetic co-simulation approach is the major factor that determines the energy estimation time required by the proposed energy estimation technique. It varies depending on the hardware-software mapping and scheduling of the tasks that constitute the application. This is due to two main reasons. One reason is due to the differences in simulation speeds of the hardware simulator and the software simulator. Table 5.7 shows the simulation speeds of the cycle-accurate MicroBlaze instruction set simulator, the MATLAB/Simulink simulation environment for simulating the customized hardware peripherals and ModelSim for timing based low-level simulation. Cycle-accurate simulation of software executions is much faster (more than 4 times faster) than cycle-accurate arithmetic level simulation of hardware execution using MATLAB/Simulink. Thus, if more tasks are mapped to execute on the customized hardware peripherals, it will slow down the overall simulation speed that can be achieved by the proposed arithmetic level co-simulation approach. Compared with the low-level simulation using ModelSim, our MATLAB/Simulink based implementation of the co-simulation approach can potentially achieve simulation speed-ups from 29.0x to more than 114x. Another reason is the frequency of data exchanges between the software program and the hardware peripherals. Every time that the simulation data is exchanged between the hardware simulator and the software simulator, the simulation processes performed within the simulators are stalled and resumed. This would add quite some extra overhead to the co-simulation process. There are close interactions between the hardware and software execution for the two numerical computation applications considered in this chapter. Thus, the speed-ups achieved for the two applications are smaller than the maximum speed-ups that can be achieved in principal.

If we further consider the time for implementing (including synthesizing and placing-and-routing) the complete system and generating the post place-and-route simulation models, which is required by the low-level energy estimation approaches, our arithmetic level co-simulation approach would lead to even much greater simulation speed-ups. For the two numerical computation applications, the time for implementing the complete system and generating the post place-and-route simulation models takes around 3 hours. Thus, our arithmetic level simulation based energy estimation technique can be 197x to 6534x faster than that based on low-level simulation for these two numerical computation applications.

For the hardware peripheral of the CORDIC division application, our energy estimation is based on the energy functions for the processing elements shown in Figure 5.33. For the hardware peripheral of the matrix multiplication application, our energy estimation is based on the energy functions for the multipliers and the accumulators. As one input to these energy functions, we calculate the average switching activity of all the input/output ports of the Simulink blocks during the arithmetic level simulation process. Table 5.6 shows the energy estimates obtained using our arithmetic level simulation based energy estimation technique. Energy estimation errors ranging from 9.5% to 18.2% and 11.6% on average are achieved for these two numerical computation applications compared with actual measurements. Low-level simulation based energy estimation using XPower achieves an average estimation error of 6.8% compared with actual measurements.

To sum it up, for the two numerical computation applications, our arithmetic level co-simulation based energy estimation technique sacrifices an average 4.8% estimation accuracy while achieving estimation speed-ups up to 6534x compared with low-level energy estimation techniques. The implementations of the two applications identified using our energy estimation technique achieve energy reductions up to 52.2% compared with other implementations considered in our experiments.

5.8   Summary

A two-step rapid energy estimation technique for hardware-software co-design using FPGAs was proposed in this chapter. An implementation of the proposed energy estimation technique based on MATLAB/Simulink and the design of two numerical computation applications were provided to demonstrate the effectiveness of our approach.

One extension of our rapid energy estimation approach is to provide confidence interval information of energy estimates obtained using the techniques proposed in this chapter. Providing such information is desired in the development of many practical systems. As the use of multiple clocks with operating frequencies is becoming popular in the development of configurable multi-processor platforms for improving time and energy performance, we extend our work to support energy estimation for hardware-software co-designs using multiple clocks. This requires that we extend the co-simulation environment to maintain and synchronize multiple global simulation timers used to keep track of the hardware and software simulation processes driven by different clock signals.

Chapter 6

Hardware-Software Co-Design for Energy Efficient Implementations of Operating Systems

6.1 Introduction

There is a strong trend towards integrating FPGA with various heterogeneous hardware components, such as RISC processors, embedded multipliers and memory blocks (e.g., Block RAMs in Xilinx FPGAs). Such integration leads to a reconfigurable System-on-Chip (SoC) platform, an attractive choice for implementing many embedded systems. *FPGA based soft processors*, which are RISC processors realized using the configurable resources available on FPGA devices, are also becoming popular. Examples of such processors include Nios from Altera [3] and MicroBlaze and PicoBlaze from Xilinx [97]. One advantage of designing using soft processors is that they provide new design trade-offs by time sharing the limited hardware resources (e.g., configurable logic blocks on Xilinx FPGAs) available on the devices. Many control and management functionalities as well as computations with tightly coupled data dependency between computation steps (e.g., many recursive algorithms such as Levinson Durbin algorithm [42]) are inherently more suitable for software implementations on processors than the corresponding customized (parallel) hardware implementations. Their software implementations are more compact and require a much smaller amount of hardware resources. Such compact designs using soft processors can effectively reduce the static energy dissipation of the complete system by fitting into smaller FPGA devices [93]. Most importantly, soft processors are "configurable" by allowing the customization of the instruction set and/or the attachment of customized hardware peripherals in order to speed up the computation of algorithms with a large degree of parallelism and/or to efficiently perform some auxiliary management functionalities as described in this chapter. The Nios processor allows users to customize up to five instructions. The MicroBlaze processor supports various dedicated communication interfaces for attaching customized hardware peripherals to it.

Real-time operating systems (RTOSs) simplify the application development process by decompling the application code into a set of separate tasks. They provide many effective mechanisms for task and interrupt management. New functions can be added to the RTOS without requiring major changes to the software. Most importantly, for preemptive real-time operating systems, when low priority tasks are added to the system, their impact on the responsiveness of the system can be minimized as much as possible. By doing so, time-critical tasks and interrupts can be handled quickly and efficiently. Therefore, real-time operating systems are widely adopted in the development of many embedded applications using soft processors.

Energy efficiency is a key performance metric in the design of many embedded systems. FPGAs provide superior performance to general-purpose processors and DSPs in terms of computation capability per power consumption [18] and have been deployed in many portable devices (e.g., hand-held camcorders [114] and software-defined radio [113]) and automotive systems [111]. Power efficiency is a crucial performance metric in the development of these usually battery operated embedded systems due to the limited power supply.

Novel technologies have been proposed and integrated into modern FPGAs that can effectively reduce both static and dynamic power consumption. The triple oxide process technology and the 6-input look-up table (LUT) architecture in the Xilinx 65 nm Virtex-5 FPGAs lead to 40% reduction in static power consumption compared with the prior generation of FPGAs [106]. The 6-input LUT architecture along with a symmetric routing structure in Virtex-5 FPGAs result in 35% to 40% core power reduction compared with the prior Virtex-4 FPGAs. Dual-V_{DD} low-power FPGA architectures have been proposed by both Gayasen, et al. [33] and Li, et al. [59]. An overall power reduction of 61% on a multimedia testbench is reported by employing the dual-V_{DD} technology. Power efficiency placement and implementation algorithms can further reduce the FPGA power dissipation [34].

Modern FPGAs offer many on-chip energy management mechanisms to improve the energy efficiency of designs using them. One important mechanism is "clock gating". The user can dynamically stop the distribution of clock signals to some specific components when the operations of these components are not required. Another important mechanism is the dynamic switching of clock sources. Many FPGA devices provide multiple clock sources with different operating frequencies and clock phases. When communicating with low-speed off-chip peripherals, portions of the FPGA device can dynamically switch to operate with a slow clock frequency and thus reduce the energy dissipation of the system. For example, Xilinx FPGAs provide BUFGCEs to realize "clock gating" and BUFGMUXs to realize dynamic switch of clock sources [97]. These BUFGCEs and BUFGMUXs automatically avoid the glitches in the clock distribution network when changing the status of the network. Customization of the FPGA based soft processor to incorporate these on-chip energy management mechanisms is crucial for improvement of energy efficiency when implementing real-time operating systems on it. The Virtex-5 FPGAs contain up to 12 digital clock management (DCM) units, in addition to numerous clock buffers and multiplexers, which can independently drive specific regions on the FPGA device depending on specific application requirements. The Xilinx Spartan-DSP devices provide suspend mode and hibernate mode, which yield at least 40% and 99% static power reduction respectively when they are not processing data.

In this chapter, we focus on the following key design problem. The FPGA device is configured with one soft processor and several on-chip customized hardware peripherals. The processor and the hardware peripherals communicate with each other through some specific bus interfaces. There are a set of tasks to be run on the soft processor based system. Each of the tasks is given as a C program. Software drivers for controlling the hardware peripherals are provided. Tasks can invoke the corresponding hardware peripherals through these software drivers in order to perform some I/O operations or to speed up the execution of some specific computations. The execution of a task may involve only a portion of on-chip hardware resources. For example, the execution of a task may require the processor, a specific hardware peripheral and the bus interface between them while the execution of another task may only require the processor, the memory interface controllers and the memory blocks that store the program instructions and data for this task. Tasks are either executed periodically or invoked by some external interrupts. Tasks have different execution priorities and are scheduled in a preemptive manner. Tasks with higher priorities are always selected for execution and can preempt the execution of tasks with lower priorities. Besides, the rates at which the tasks are executed is considered to be "infrequent". That is, there are intervals during which no tasks are under execution. Based on these assumptions, our objective is *to perform customization of the soft processor and adapt it to the real-time operating system running on it so that the energy dissipation of the complete system is minimized*. We focus on the utilization of on-chip programmable resources and energy management mechanisms to achieve our objective. Also, it is desired that the required changes to the software source code of the operating systems are kept minimal when applying the energy management techniques.

To address the design problem described above, we propose a technique based on hardware-software co-design for energy efficient implementation of real-time operating systems on soft processors. The basic ideas of our approach are the integration of on-chip energy management mechanism and the employment of customized "hardware" assisted task management component by utilizing the configurability offered by soft processors. More specifically, we tightly couple several customized hardware components to the soft processor, which perform the following functionalities for energy management: (1) manage the clock sources for driving the soft processor, the hardware peripherals, and the bus interfaces between them; (2) perform the task and interrupt management responsibilities of the operating system cooperatively with the processor; (3) selectively wake up the processor and the corresponding hardware components for task execution based on the hardware resource requirements of the tasks. Since the attached hardware peripherals can work concurrently with the soft processor, the proposed hardware-software co-design techniques incur an extra clock cycle of task and interrupt management overhead, which is negligible for the development of many embedded systems compared with the software management overhead. Most importantly, due to the parallel processing capability of the hardware peripherals, we show that for some real-time operating systems, our techniques can actually lead to less management overheads compared with the corresponding "pure" software implementations. In addition, we implement a real-time operating system on a state-of-the-art soft processor to illustrate our approach. The development of several embedded applications is provided in both Section 6.6 and Section 6.7 to demonstrate the effectiveness of our approach. Actual measurement on an FPGA prototyping board shows that the systems implemented using our power management techniques achieve energy reductions ranging from 73.3% to 89.9% and 86.8% on the average for the different execution scenarios of the applications considered in our experiments.

This chapter is organized as follows. Section 6.2 discusses the background information about real-time operating systems as well as the related work about customization of FPGA based soft processors. Section 6.5 presents our hardware-software co-design technique for energy efficient implementation of real-time operating systems on FPGA based soft processors. The implementations of two popular operating systems through customization of a state-of-the-art soft processor are described in Section 6.6 and Section 6.7 in order to illustrate our energy management techniques. The development of two embedded FFT applications, the execution of which is typical in many embedded systems, are provided in both of these two sections to demonstrate the effectiveness of our approach. Finally, we conclude in Section 6.8.

6.2 Real-Time Operating Systems

6.2.1 Background

Real-time operating systems may encompass a wide range of different characteristics. We focus on the typical characteristics listed below in this chapter. We retain these important features of real-time operating systems when applying our energy management techniques to improve their energy efficiency.
- *Multitasking*: The operating system is able to run multiple tasks "simultaneously" through context switching. Each task is assigned a priority number. Task scheduling is based on the priority numbers of the tasks ready to run. When there are multiple tasks ready for execution, the operating system always picks up the task with the highest priority for execution.
- *Preemptive*: The task under execution may be intercepted by a task with

higher priority or an interrupt. A context switching is performed during the interception so that the intercepted task can resume execution later.

• *Deterministic*: Execution times for most of the functions and services provided by the operating system are deterministic and do not depend on the number of tasks running in the user application. Thus, the user should be able to calculate the amount of time the real-time OS takes to execute a function or a service.

• *Interrupt management*: When interrupts occur, the corresponding interrupt service routines (ISRs) are executed. If a higher priority task is awakened as a result of an interrupt, this highest priority task runs as soon as all nested interrupts are completed.

6.2.2   Off-the-Shelf Operating Systems

Several commercial real-time operating systems have been ported to soft processors. ThreadX [27], a real-time operating system from Express Logic, Inc., has been ported to both MicroBlaze and Nios processors. In ThreadX, only the system services used by the application are brought into the run-time image of the operating system. Thus, the actual size of the operating system is determined by the application running on it.

6.2.2.1   MicroC/OS-II

MicroC/OS-II is a real-time operating system that supports all the important characteristics discussed above [56]. Each task within MicroC/OS-II has a unique priority number assigned to it. The operating system can manage up to 64 tasks. There is a port of MicroC/OS-II for MicroBlaze processor. In this implementation, the operating system runs a dummy *idle* task when no useful task waits for execution and no interrupt presents. Xilinx also provides a real-time operating system for application development using MicroBlaze. However, aside from the various benefits offered by these real-time operating systems, none of these implementations of real-time operating systems addresses the energy management issue in order to improve their energy efficiency when running on FPGA based soft processors.

6.2.2.2   TinyOS

TinyOS [92] is an operating systems designed for wireless embedded sensor networks. It adopts a component-based architecture, which leads to a tighter integration between the user application code and the OS kernel than traditional operating systems. Not all the features of real-time operating systems discussed in Section 6.2.1 are supported in TinyOS. For example, it may delay the processing of tasks till the arrival of the next interrupt. The unpredictable and fluctuate wireless transmission makes such full support unnecessary. TinyOS turns the processor into a low-power sleep mode when no processing is required and wakes it up when interrupts occur. Lacking the customized hardware peripherals proposed in this chapter to take over the management responsibilities when the processor is in sleep mode, TinyOS would result in unnecessary wake-ups of the processor when processing tasks such as the periodic management of OS clock ticks.

Operating systems with a component based architecture provide a set of reusable system *components*. Each component performs some specific functionalities (e.g., OS services and FFT computation). Implementations of the components can be *pure* software programs or software wrappers around hardware components. An application designer builds an applications by connecting the components using some "wiring specifications", which are independent of the implementations of the components. Decomposing the different functionalities provided by the operating system into separate components allows unused functionalities to be excluded from the application.

nesC is a system programming language based on ANSI C language that supports the development of component based operating systems. It has been used to build several popular operating systems such as *TinyOS* [92]. It is also used to implement networked embedded systems such as *Motes* [22]. There are two types of components in *nesC*: *modules* and *configurations*. Modules provide the actual implementations of the functionalities as well as the *interfaces* with other components. Interfaces describe the interactions between different components and are the only access point to the components. Configurations are used to wire components together, connecting interfaces used by a component to interfaces of another component. An application described in *nesC* contains a *top-level configuration* that wires together the components used by it. In addition, *nesC* provides the following features to improve the time performance of component based operating systems.

• *Task and event based concurrency*: *nesC* provides two kinds of concurrency: *tasks* and *events*. Tasks are a deferred execution mechanism. Components can *post* tasks and the *post* operation immediately returns. The actual execution of the tasks is deferred until the task scheduler releases them later. Tasks run to completion and do not preempt each other. To ensure low task execution latency, individual tasks must be short. Tasks with long execution should be decomposed into multiple tasks in order to keep the system reactive. *Events* also run to completion but may preempt the execution of a task or another event. Events signify either the completion of a split-phase operation (discussed below) or the occurrence of interrupts (e.g., message reception or timer time-out). The simple concurrency model of *nesC* allows high concurrency with low overhead. This is compared against the thread-based concurrency model used by the operating systems discussed in Section 6.2.2, where the thread stacks consume precious memory space while blocking incoming processing requests.

• *Split-phase operations*: Since tasks in *nesC* run non-preemptively, the operations of the tasks with long latency are executed in a *split-phase* manner. Interfaces in *nesC* are *bi-directional*. One component can request the other component to perform an operation by issuing a *command* through an interface. For a *split-phase* operation, the commands return immediately. The component can be notified of the completion of the operation by implementing the corresponding *event* for the operation through the same interface.

• *Whole-program optimization*: According to the wiring of the components that constitute the application, *nesC* generates a single C program. Program analysis (i.e., data races) and performance optimization are performed based on the single C program. Such whole-program optimization leads to more compact code size and more efficient interactions between components. For example, the kernel of *TinyOS* requires a total 400 bytes of program and data memory, a significant reduction of kernel size compared with the ~15 Kbytes storage requirement of the MicroC/OS-II kernel.

See [32] and [92] for more discussions about *nesC* and *TinyOS*.

6.3 On-Chip Energy Management Mechanisms

For the commercial and academic real-time operating systems discussed above, all the task and interrupt management responsibilities of the operating systems are implemented in "pure" software. In these implementations, the operating systems have little or no control of the status of the processor and its peripherals. As is discussed in Section 6.4, even though there are already several real-time operating systems ported to soft processors, to our best knowledge, there have been no attempts to perform customized "configuration" of the soft processors and provide energy management capability to the operating systems in order to improve their energy efficiency. The dynamic voltage and frequency scaling (DVFS) technique has proved to be an effective method of achieving low power consumption while meeting the performance requirements [45]. DVFS is employed by many real-time operating systems. However, DVFS cannot be realized directly on the current reconfigurable SoC platforms and requires additional and relatively sophisticated off-chip hardware support. Thus, application of DVFS is out of scope for the design problem discussed in this chapter. Another effective energy management technique is to power off the processors when no tasks are ready for execution. This technique has been used in the design of operating systems such as TinyOS [92]. In these designs, the hardware component that controls the powering on/off of the processors is loosely coupled with the operating systems and share little or no responsibilities of the operating system with the processor. For example, the OS clock clicks can only be managed by the processor for the design of MicroC/OS-II shown in [56]. This would result in unnecessary waking up of the processor and undesired energy dissipation in many cases. For TinyOS, such powering on and off may also cause delays for executing tasks with high priority. Selective waking up of the hardware components depending on the resource requirement of the tasks is not realized in these designs.

6.4 Related Work

There have been quite some efforts on customizing soft processors in order to optimize their performance toward a set of target applications. Cong et al. propose a technique based on shadow registers so as to get a better utilization of the limited data bandwidth between the soft processor and the tightly coupled hardware peripherals [19]. In their technique, the core register file of the soft processor is augmented by an extra set of shadow registers which are conditionally written by the soft processor in the write-back stage and are read only by the hardware peripherals attached to the processor.

Shannon and Chow propose a programmable controller with a SIMPPL (Systems Integrating Modules with Predefined Physical Links) system computing model as a flexible interface for integrating various on-chip computing elements [83]. The programmable controller allows the addition of customized hardware peripherals as computing elements. All the computing elements within the controller can communicate with each other through a fixed physical interface. Their approach enables users to easily adapt the controller to the new computing requirements without necessitating the redesign of other elements in the system.

Ye et al. propose the optimization technique based on run-time reconfiguration units to adapt the processor to the ever-changing computing requirements [117]. Sun et al. propose a scalable synthesis methodology for customizing the instruction set of the processor based on the specific set of application of interests [88]. Hubner et al. [118] have a design based on multiple soft processors for automotive applications. In their designs, each of the multiple soft processors is optimized to perform some specific management responsibilities.

Shalon and Mooney propose a System-on-Chip dynamic memory management unit (SoCDMMU) for managing the global memory shared between multiple on-chip processors [82]. They have modified an existing RTOS to support the SoCDMMU unit. They extend their work and propose a more generic hardware-software RTOS framework for SoC platforms and FPGAs [65]. A user can choose to implement some of the OS functionalities in either software or hardware using their framework, and a customized operating system is generated based on the user's selection. Significant execution speed-ups are reported using their SoCDMMU unit and the co-design framework. Nevertheless, their research does not focus on improving the power efficiency of RTOSs, which is the major contribution of our work.

In spite of all the previous work on soft processors, to our best knowledge, the hardware-software co-design technique proposed in this chapter is the first attempt to customize a soft processor to optimize the energy performance of the real-time operating system running on it.

6.5 Our Approach

The basic idea of our hardware-software co-design based technique for energy efficient implementation of real-time operating system is to tightly couple several dedicated hardware peripherals to the soft processor. These hardware peripherals cooperatively perform task and interrupt management responsibilities of the operating system together with the software portions of the operating system running on the soft processor. These hardware peripherals also control the activation states of the various on-chip hardware components including the processor by utilizing the on-chip energy management mechanisms. They take over the task and interrupt management responsibilities of the operating systems running on the soft processor when there is no task ready for execution. In this case, except for these energy management components, all the other hardware components including the soft processor are completely shut off when they are not processing any "useful" tasks. There are two major reasons for improvement of energy efficiency using our hardware-software co-design technique. One reason is that the energy dissipation of these attached dedicated hardware components for performing task and interrupt management are much smaller than that of the soft processor. Another reason is that we selectively wake up the hardware peripherals attached to the processor and put them into proper activation states based on the hardware resource requirements of the tasks under execution. Thus, the undesired energy dissipation of the hardware peripherals that are not required for execution can be eliminated and the energy efficiency of the complete system can be further improved.

The overall hardware architecture of our hardware-software co-design approach is shown in Figure 6.1. Three hardware components, which are clock management unit, auxiliary task and interrupt management unit, and selective component wake-up unit, are added to the soft processor to perform energy management.

• *Clock management unit*: Many FPGA devices provide different clock sources on a single chip. The FPGA devices can then be divided into different clock domains. Each of the clock domains can be driven by different clock sources. Dynamic switching between the clock sources with different operating frequencies within a few clock cycles is also possible. For example, when the processor is communicating with a low-speed peripheral, instead of running the processor in some dummy software loops when waiting for the response

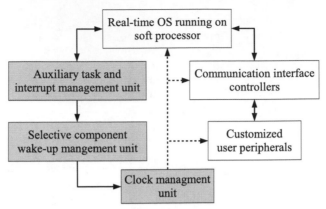

---- denotes the transmission of clock signals

FIGURE 6.1:   Overall hardware architecture

from slow peripherals, the user can choose to switch the processor and the related hardware peripherals to work with a clock source with a slower operating frequency. This can effectively reduce the energy dissipated by the processor. Clock gating is another important technique used in FPGA designs to reduce energy dissipation. The user can dynamically change the distribution of clock signals and disable the transmission of clock signals to the hardware components that are not in use.

The clock management unit provides explicit control access to the clock distribution network of the FPGA device. It accepts the control signals from other components and change clock sources that drive the various hardware components using the two techniques discussed above.

• *Auxiliary task and interrupt management unit*: We attach an auxiliary task and interrupt management (ATIM) hardware peripheral to the soft processor. We let the internal data of the operating system for task scheduling and interrupt management be shared between the ATIM unit and soft processor. Hence, when the operating system determines that no task is ready for execution and no interrupt presents, the ATIM unit sends out signals to the clock management component unit and disables the transmission of clock signals to the soft processor and the other hardware peripherals attached to it. The ATIM takes over the task and interrupt management responsibilities of the operating system. When the ATIM unit determines that a task is ready for execution or any external interrupt arrives, it wakes up the processor and the related hardware peripherals and hands the task and interrupt management responsibilities back to the processor. Note that to retain the deterministic feature of the real-time operating system, dedicated bus interfaces with deterministic delays are used for communication between the processor and the other hardware components.

• *Selective component wake-up and activation state management unit*: Since the execution of a task uses only a portion of the device, we provide a selective wake-up state management mechanism within the OS kernel. Using the clock management unit, the FPGA device is divided into several clock domains. Each of the clock domains can be in either "active" or "inactive" (clock gated) state. We denote a combination of the activation states of these different clock domains as an *activation state* of the device. It is the user's responsibility that a task is assigned to an appropriate activation state of the device in which the hardware components required by the task are all active. Thus, when a task is selected by the operating system for execution, only some specific components used by the task are driven by the clock sources. The unused components are kept in clock gated state in order to save power consumption.

In the following sections, we implement two state-of-the-art real-time operating systems on the MicroBlaze soft processor. We perform customization to the MicroBlaze processor and enhance the energy performance of the two real-time operating systems by applying the hardware-software co-design energy management technique proposed in this chapter.

Regarding the management overhead caused by our hardware-software co-design techniques, since the attached hardware peripherals can work concurrently with the soft processor, the proposed hardware-software co-design techniques incur an extra clock cycle of task and interrupt management overhead, which is negligible for the development of many embedded systems compared with the software management overhead. Most importantly, due to the parallel processing capability of the hardware peripherals, we show that for some real-time operating systems (e.g., *h-TinyOS* discussed in Section 6.7), our techniques can actually lead to less management overheads compared with the corresponding "pure" software implementations.

6.6 An Implementation Based on MicroC/OS-II

The hardware architecture of the complete MicroC/OS-II operating system is shown in Figure 6.2. Except for the "priority aliasing" technique discussed in Section 6.6.4, our energy management techniques are transparent to the software portions of the operating system and do not require changes in software.

6.6.1 Customization of MicroBlaze Soft Processor

The instruction and program data of the MicroC/OS-II operating system are stored at the BRAMs. The MicroBlaze processor gets access to the instruction and the program data through two LMB (Local Memory Bus) interface controllers, one for instruction side and the other for data side data access.

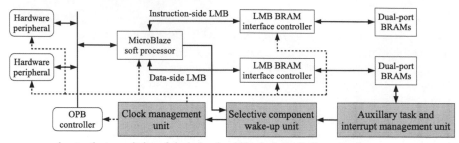

---- denotes the transmission of clock signals within clock distribution network

FIGURE 6.2:	Configuration of the MicroBlaze soft processor with the COMA scheme

There are two different kinds of bus interfaces to attach customized hardware peripherals to MicroBlaze. One kind of bus interface is Faster Simplex Links (FSLs), which are dedicated interfaces for tightly coupling high speed peripherals to MicroBlaze. A 32-bit data can be sent between the processor and its hardware peripherals through FSLs in a fixed time of two clock cycles. To retain the real-time responsiveness of the operating system, FSL interfaces are used for attaching the energy management hardware peripherals to the MicroBlaze soft processor. Another interface is the On-chip Peripheral Bus (OPB) interface, a shared bus interface for attaching hardware peripherals that operate with a relative low frequency compared to the MicroBlaze processor. For example, peripherals for accessing general purpose input/output (GPIO) interfaces and those for managing the communication through serial ports can be attached to MicroBlaze through the OPB bus interface.

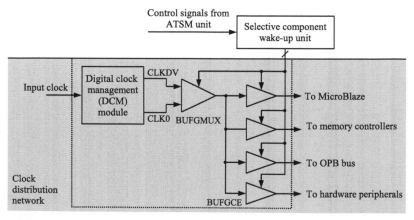

FIGURE 6.3:	An implementation of the clock management unit

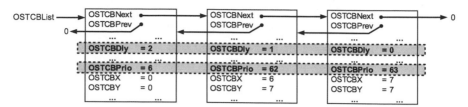

FIGURE 6.4: Linked list of task control blocks

6.6.2 Clock Management Unit

The hardware architecture of the clock management unit is illustrated in Figure 6.3. Xilinx Spartan-3/Virtex-II/Virtex-II Pro/Virtex FPGAs integrate on-chip digital clock management (DCM) modules. Each DCM module can provide different clock sources (CLK0, CLK2X, CLKDV, CLKFX, etc.). Each of these clock sources can be used to form a clock domain and drive the hardware components within its own domain. Thus, different on-chip hardware components can operate under different operating frequencies. For example, on Spartan-3 FPGA devices, CLKDV of the DCM module can divide the input clock by up to 16. For the design examples shown in Section 6.6 and Section 6.7, when the input clock frequency is 50 MHz, the output clock frequency of CLKDV can be as low as 3.125 MHz.

There are multiplexers (i.e., BUFGMUXs) within the clock distribution network. BUFGMUXs can be used to dynamically switch the clock sources with different operating frequencies for driving the soft processor, other hardware peripherals, and the communication interfaces between them. Xilinx FPGAs also integrate buffers with enable port (BUFGCEs) within their clock distribution network which can be used to realize "clock gating". That is, these BUFGCEs can be used to dynamically drive the corresponding hardware components only when these components are required for execution.

The clock management unit accepts control signals from the selective component wake-up and activation state management unit (discussed in Section 6.6.4) and changes the clock sources for driving the soft processor, the memory controllers, the OPB bus and the FSLs connected to the soft processor accordingly. Using BUFGMUXs and BUFGCEs, the clock management unit can change the activation state of the FPGA device within one or two clock cycles upon receiving the request from the other management hardware components. As is analyzed in Section 6.6.5, by comparing the software overhead for context switching between tasks, the addition of the clock management unit introduces negligible overhead to the operating system.

6.6.3 Auxiliary Task and Interrupt Management Unit

In order to take over the responsibilities of the operating systems when the processor is turned off, the auxiliary task and interrupt management unit performs three major functionalities: ready task list management, OS clock tick management and interrupt management. They are described in detail below.

• **Ready task list management**: MicroC/OS-II maintains a ready list consisting of two variables, OSRdyGrp and OSRdyTbl to keep track of the task status. Each task is assigned a unique priority number between 0 and 63. As shown in Figure 6.5, tasks are grouped (eight tasks per group) and each group is represented by one bit in OSRdyGrp. OSRdyGrp and OSRdyTbl actually form a table. Each slot in the table represents a task with a specific priority. If the value of a slot equals zero, it means that the task represented by this slot is not ready for execution. Otherwise, if the value of the slot is one, it means that the task represented by this slot is ready for execution. Whenever there are tasks ready for execution, the operating system searches the two variables in the ready list table to find out the task with highest priority that is ready to run. A context switch is performed if the selected task has higher priority than the task under execution. As mentioned in Section 6.2.2, MicroC/OS-II always runs an idle task (represented by slot 63) with the lowest priority when no other task is ready for execution.

We use some complier construct to explicitly specify the storage location of the ready task list (i.e., OSRdyGrp and OSRdyTbl) on the dual-port BRAMs. The MicroBlaze processor can get access to OSRdyGrp and OSRdyTbl through the data side LMB bus controller attached to port A of the dual-port BRAMs while the ATIM component can also get access to these two variables through port B of the BRAMs. The ATIM component keeps track of the two variables of the ready list with a user-defined period. When it detects that only the idle task is ready for execution, it will signal the clock management unit to disable the clocks sent to the processor, its hardware peripherals, and the bus interfaces between them. These "clock gated" components will resume their normal states when useful tasks become ready for execution and/or external interrupts are presented.

• **OS clock tick management**: OS clock ticks are a special kind of interrupts that are used by MicroC/OS-II to keep track of the time experienced by the system. A dedicated timer is attached to the MicroBlaze processor and repeatedly generates time-out interrupts with a pre-defined interval. Each interrupt generated by this timer corresponds to one clock tick. MicroC/OS-II maintains an 8-bit counter for counting the number of clock ticks experienced by the operating system. Whenever the MicroBlaze processor receives a time-out interrupt from the timer, MicroC/OS-II increases the clock tick counter by one. The counting of clock ticks is used to keep track of time delays and time-outs. For example, the period that a task is repeatedly executed is counted in clock ticks. Also, a task can stop waiting for an interrupt if the interrupt of interest fails to occur within a certain amount of clock ticks. We use a special hardware component to perform the OS clock tick management and separate it from the management of interrupts for other components through the OPB bus. By doing so, frequently powering on and off of the OPB bus controller

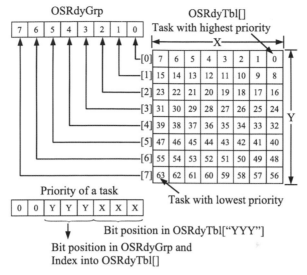

FIGURE 6.5:   Ready task list

to notify the processor of the clock tick interrupt, which is required by other interrupts and would result in unnecessary energy dissipation, can be avoided.

When a task is created, it is assigned a *task control block*, OS_TCB. A task control block is a data structure used by MicroC/OS-II to maintain the state of a task when it is preempted. The task control blocks are organized as a linked list, which is shown in Figure 6.4.

When the task regains controls of the CPU, the task control block allows the task to resume execution from the previous state where it has stopped. Specifically, the task control block contains a field OSTCBDly, which is used when a task needs to be delayed for a certain number of clock ticks or a task needs to pend for an interrupt to occur within a timeout. In this case, OSTCBDly field contains the number of clock ticks the task is allowed to wait for the interrupt to occur. When this variable is 0, the task is not delayed or has no timeout when waiting for an interrupt. MicroC/OS-II requires a periodic time source. Whenever the time source times out, the operating system decreases OSTCBDly by 1 till OSTCBDly = 0, which indicates that the corresponding task is ready for execution. The bits in OSRdyGrp and OSRdyTbl representing this task will then be set accordingly.

In order to apply the proposed energy management technique, the OSTCBDly field for each task is stored at a specific location of the dual-port BRAMs. The MicroBlaze processor gets access to OSTCBDly through the data side LMB bus controller attached to port A of the dual-port BRAMs. On the other hand, the ATIM component can also get access to OSTCBDly through port B of the BRAMs.

When the MicroBlaze processor is turned off by the clock management unit,

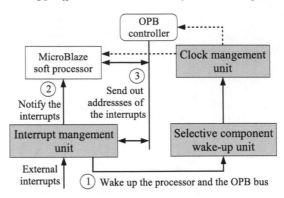

FIGURE 6.6:   Interrupt management unit

the ATIM unit will take over the management responsibilities of MicroC/OS-II from the processor. The ATIM unit gets access to the OSTCBDly fields of the task control blocks through another port of the dual-port BRAMs and decreases their values when an OS clock tick interrupt occurs. Whenever the OSTCBDly field of some tasks reaches zero, which means that these tasks are ready for execution, the ATIM unit will signal the clock management unit, which will bring up the processor together with the related hardware components to process the tasks that are woken up and ready for execution.

• **Interrupt management**: The architecture of the interrupt management unit is shown in Figure 6.6. When external interrupts arrive, the interrupt management will check the status of the MicroBlaze processor and the OPB bus controller. If neither the processor nor the OPB bus controller are active, the interrupt management unit will perform the following operations: (1) send out control signals to the selective component wake-up unit to enable the processor and the OPB controller; (2) notify the processor of the external interrupts; and (3) when the processor responds and starts querying, send out the interrupt vector through the OPB bus and notify the soft processor of the sources of the incoming external interrupts. If both the processor and the OPB bus controller are already in active state, the interrupt management unit will skip the wake-up operation and go on directly with operation (2) and (3) described above.

6.6.4   Selective Wake-up and Activation State Management Unit

To minimize the required changes to the software portion of the MicroC/OS-II operating system, we employ a technique called "priority aliasing" to realize the selective component wake-up and activation state management unit. As shown in Figure 6.5, MicroC/OS-II uses an 8-bit unsigned integer to store the priority of a task. Since MicroC/OS-II supports a maximum number of

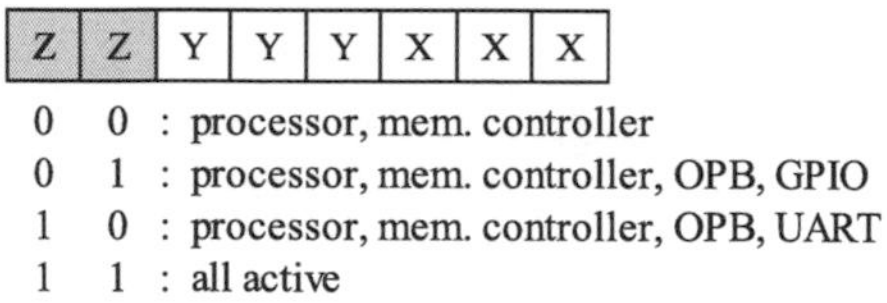

Z	Z	Y	Y	Y	X	X	X

0   0 : processor, mem. controller
0   1 : processor, mem. controller, OPB, GPIO
1   0 : processor, mem. controller, OPB, UART
1   1 : all active

FIGURE 6.7:   Priority aliasing for selective component wake-up and activation state management

64 tasks, only the last 6 bits of the unsigned integer number is used to denote the priority of a task. When applying the "priority aliasing" technique, we use the first 2 bits of a task's priority to denote the four different combinations of activation states of the different hardware components (i.e., an activation state of the FPGA device). Thus, the user can assign a task with a specific priority using four different task priority numbers. Each of these four numbers corresponds to one activation state of the FPGA device. It is up to the user to assign an appropriate activation state of the device to a task. One possible setting of the two-bit information for the design example discussed in Section 6.6.6 is shown in Figure 6.7. The user can choose to support other activation states of the FPGA device by making appropriate changes to the clock management unit and the selective component wake-up management unit.

Note that when applying the "priority aliasing" technique, the processor can also change the activation state of the device during its normal operation depending on the tasks that are under execution. As shown in Figure 6.4, OSTCBPrio in the task control blocks are stored at some specific locations on the dual-port BRAMs and are accessible to both the auxiliary task and interrupt management unit and the MicroBlaze processor through the dual-port memory blocks on the FPGA device. When MicroBlaze is active, it can send the first two bits of the task priority numbers that determine the activation states of the device to the selective wake-up management unit through an FSL bus interface before it starts executing a task.

6.6.5   Analysis of Management Overhead

Figure 6.8 shows the typical context switch overhead of the MicroC/OS-II operating system. Task A finishes its processing and calls OSTimeDly() to stop its execution, which will resume till a later time. Then, a context switch is performed by MicroC/OS-II and task B is selected for execution as a result of the context switch. During the function call OSCtxSw() for the actual context switch operations (which include saving the context of task A and then restoring the context of task B), MicroC/OS-II uses one clock cycle to issue a request to the selective component wake-up management unit. The request will be processed by both the selective component wake-up management unit and the clock management unit to change the activation state of the FPGA

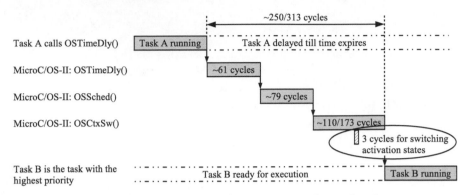

FIGURE 6.8: Typical context switch overhead

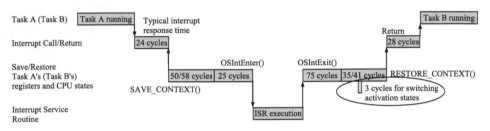

FIGURE 6.9: Typical interrupt overhead

device to the state that is specified by task B. Note that the change of activation states performed by the selective component wake-up management unit and the clock management unit is in parallel with the other required context switching operations performed by MicroC/OS-II on the MicroBlaze processor. Therefore, while the typical context switch is between 250 to 313 clock cycles, the proposed hardware-software co-design technique actually incurs one extra management clock cycle overhead.

Figure 6.9 illustrates the typical interrupt overhead of the MicroC/OS-II operating system. Task A is under execution and an external interrupt occurs. MicroC/OS-II responds to the incoming interrupt and executes the corresponding interrupt service routine (ISR). Task B, which has a higher priority than task A, is made ready for execution as a result of executing the ISR. Since task B is running at a different activation state as that of task A, a change in the activation states of the FPGA device is required when MicroC/OS-II is restoring the execution context for task B. Similar to the context switch case discussed above, MicroC/OS-II spends one extra clock cycle during the restoration of context in order to issue the request for changing the activation state to the selective component wake-up management unit and the clock management unit. The next 3 clock cycles of actual change of activation states are performed in the selective component wake-up management unit and the clock management unit in parallel with MicroC/OS-II, which is still in the process of restoring the context for executing task B.

To summarize, based on the analysis of the above two cases, we can see that the management overhead incurred by the proposed hardware-software co-design technique is only one clock cycle and is negligible in many practical application developments.

6.6.6   Illustrative Application Development

In this section, we show the development of an FFT computation application and using the MicroC/OS-II operating system running on a MicroBlaze processor enhanced with the proposed hardware-software co-design technique. FFT is a widely deployed application. The scenarios of task execution and interrupt arrivals considered in our examples are typical in the design of many embedded systems, such as radar and sensor network systems and software defined radio, where energy efficiency is a critical performance metric. Actual measurements on a commercial FPGA prototyping board is performed to demonstrate the effectiveness of our hardware-software co-design approach.

6.6.6.1   Customization of the MicroBlaze Soft Processor

The configuration of the MicroBlaze soft processor is as shown in Figure 6.2. Two customized hardware peripherals are attached to MicroBlaze through the OPB bus interface. One peripheral is an opb_gpio hardware peripheral that accepts data coming from an 8-bit GPIO interface. The other one is an opb_uartlite hardware peripheral that communicates with an external computer through serial UART (Universal Asynchronous Receiver-Transmitter) protocol.

Based on the configuration discussed above, the settings of the selective component wake-up and activation state management unit is shown in Figure 6.7. The first two bits of a task priority denote four different activation states of the device. "00" represents the activation state of the device when only the processor and the two LMB bus controllers are active. Under the activation state, the processor can execute tasks for which the instructions and data are stored in the BRAMs accessible through the LMB bus controllers. Access to other peripherals is not allowed in the "00" activation state. The experimental results shown in Table 6.1 indicate that the "00" status reduces the power consumption of the device by 23.6% compared with activation state "11" when all the hardware components are active. The device has two other activation states: "01" stands for the state that the processor, the two memory controllers, the OPB bus controller, and the general purpose I/O hardware peripherals are active; "10" stands for the state that the processor, the two memory controllers, the OPB bus controller, and the hardware peripheral for communication through the serial port are active. As shown in Figure 6.3, in states "00", "01" and "11", the FPGA device is driven by CLK0 while in state

"10", it is driven by CLKDV, the operating frequency of which is 8 times less than that of CLK0.

6.6.6.2 A Case Study

To demonstrate the effectiveness of our hardware-software co-design technique, we developed an FFT computation application using the MicroC/OS-II real-time operating system running on the MicroBlaze soft processor.

• **Implementation:** The FFT computation application consists of three tasks: the *data-input* task, the *FFT computation* task, and the *data-output* task. The *data-input* task is responsible for storing the input data coming from the 8-bit opb_gpio hardware peripheral to the on-chip BRAM memory blocks. The *data-input* task is woken up as a result of the external interrupt from the general purpose input/output ports (GPIOs) when the data become available. One task is for FFT computation while the other task is for sending out data to the external computer for displaying through the opb_uartlite peripheral. The FFT computation task performs a 16-point complex number FFT computation as described in [77]. We consider int data type. The cos() and sin() functions are realized through table look-up. When input data presents, the MicroBlaze processor receives an interrupt from the 8-bit GPIO. MicroC/OS-II running on MicroBlaze will execute the interrupt service routine (ISR), which stores the data coming from the 8-bit GPIO at the BRAMs, and mark the FFT computation task to be ready in the task ready list. Then, after the data input ISR completes, MicroC/OS-II will begin processing the FFT computation task. The data output task is executed repeatedly with a user-defined interval. Through an opb_uartlite hardware peripheral which controls the RS-232 serial port, the data output task sends out the results of the FFT computation task to a computer where the results get displayed. Each of the 8-bit data is sent out in a fixed 0.05 msec interval. MicroBlaze runs an empty *for* loop between the transmission intervals.

To better demonstrate the effectiveness of our energy management techniques, we generate the input data periodically and vary the input and output data rates in our experiments to show the average power reductions achieved when applying our energy management techniques.

For the experiments discussed in this chapter, the MicroBlaze processor is configured on a Xilinx Spartan-3 xc3s400 FPGA [97]. The input clock frequency to the FPGA device is 50 MHz. The MicroBlaze processor, the two LMB interface controllers as well as the other hardware components shown in Figure 6.2 are operating at the same clock frequency. An on-chip digital clock management (DCM) module is used to generate clock sources with different operating frequencies ranging from 6.25 MHz to 50 MHz for driving these hardware components. We use EDK 6.3.02 for describing the software execution platform and for compiling the software programs. ISE 6.3.02 [97] is used for synthesis and implementation of the complete applications. Actual power consumption measurement of the applications is performed using

TABLE 6.1: Dynamic power consumption of the FPGA device in different activation states

State	Power (W)	Reduction*	Note
00	0.212	57.1%	@ 50 MHz
01	0.464	6.1%	@ 50 MHz
10	0.026	94.7%	@ 6.25 MHz
11	0.494	–	@ 50 MHz

∗: Power reduction is compared against that of state 11.

a Spartan-3 prototyping board from Nu Horizons [66] and a SourceMeter 2400 from Keithley [53]. We compare the differences in power consumption of the FPGA device when MicroC/OS-II is operated with different activation states. We ensure that except for the Spartan-3 FPGA chip, all the other components on the prototyping board (e.g., the power supply indicator and the SRAM chip) are kept in the same operating state when the FPGA device is executing under different activation states. Under these settings, we consider that the changes in power consumption of the FPGA prototyping board are mainly caused by the FPGA chip. Using the Keithley SourceMeter, we fix the input voltage to the FPGA prototyping board at 6 Volts and measure the changes of input current to it. Dynamic power consumption of the MicroC/OS-II operating system is then calculated based on the changes of input current.

• **Experimental results:** Table 6.1 shows the power consumption of the FPGA device when MicroC/OS-II is processing the FFT computation task and the FPGA device is assigned to different activation states. Note that the measurement results shown in Table 6.1 only account for the differences in dynamic power consumption caused by these different activation states. Quiescent power, which is the power consumed by the FPGA device when there is no switching activities on it, is ignored. This is because quiescent power is fixed for a specific FPGA device and cannot be optimized using the techniques proposed in this chapter. Power reductions ranging from 6.1% to 94.7% and 52.6% on average are achieved by selectively waking up the various hardware components through the clock management unit.

Figure 6.10 shows the instant power consumption of the FPGA device when MicroC/OS-II is processing the data-input interrupt and the FFT computation task. At time interval "a", only the ATIM unit is active. All the other hardware components are inactive. At time interval "b", input data is presented. The GPIO peripheral raises an interrupt to notify the ATIM unit of the incoming of the data. Upon receiving the interrupt, the ATIM unit turns the FPGA device into activation state 01 and wakes up the MicroBlaze processor and other hardware peripherals. MicroC/OS-II begins processing the data input interrupt and stores the input data at the BRAMs through the data-side LMB bus. MicroC/OS-II also changes the ready task list and marks the FFT computation task ready for execution. Note that in order to better observe the

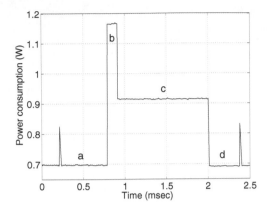

FIGURE 6.10:   Instant power consumption when processing data-input interrupt and FFT computation task

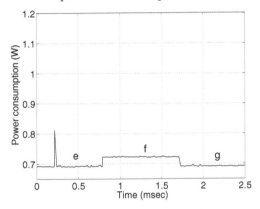

FIGURE 6.11:   Instant power consumption when processing data-output task

changes of power consumption during time interval "b", some dummy loops are added to the interrupt service routine to extend its execution time. At time interval "c", the MicroBlaze processor sends commands to the selective wake-up management unit through an FSL channel and changes the activation state of the device to state 00. Once the FPGA device enters the desired activation state, MicroC/OS-II starts executing the FFT computation task. Finally, at time interval "d", MicroC/OS-II already finishes processing the data-input interrupt and the FFT computation task. By checking the ready task list and the task control blocks, the ATIM unit detects that there are no tasks ready for execution. It then automatically disables the transmission of clock signals to all the hardware components including the MicroBlaze processor and takes over the management responsibilities originally performed by MicroC/OS-II.

In Figure 6.11, we show the instant power consumption of the FPGA de-

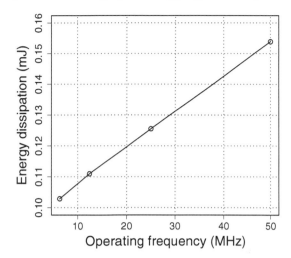

FIGURE 6.12:   Energy dissipation of the FPGA device for processing one instance of the data-out task with different operating frequencies

vice when MicroC/OS-II is processing the data-output task. At time interval "e", the ATIM unit is active and is managing the status of the tasks. The ATIM unit decreases the `OSTCBDly` fields of the task control blocks. All the other hardware components including MicroBlaze are inactive and are put into "clock gated" states. At time interval "f", the ATIM unit finds out that the `OSTCBDly` field for the data-out task reaches zero, which denotes that this task is ready for execution. The ATIM then marks the corresponding bit in the ready task list that represents this task to be 1. It also changes the FPGA device to activation state 10 according to the first two bits of the data-output task's priority number. Under activation state 10, the MicroBlaze soft processor is woken up to execute the data-out task and send out the results of the FFT computation task through the `opb_uartlite` hardware peripheral. Time interval "g" is similar to time interval "d" shown in Figure 6.10. As no tasks are ready for execution during this interval, the ATIM unit disables the other hardware components including the MicroBlaze soft processor and resumes the management responsibility of MicroC/OS-II in representation of the MicroBlaze soft processor.

There are intermittent spikes for the instant power consumption shown in Figure 6.10 and Figure 6.11. These spikes are caused by the ATIM unit when it is processing the interrupts for managing OS clock ticks. Limited by the maximum sampling rate that can be supported by the Keithley SourceMeter, we are unable to observe all the spikes caused by the OS clock tick management.

Figure 6.12 shows the power consumption of the FPGA device when MicroC/OS-II is processing the data-out task and the MicroBlaze processor is operating with different operating frequencies. Energy reduction ranging

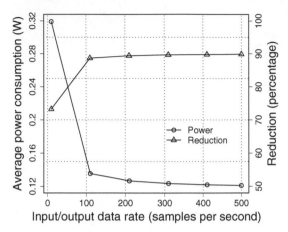

FIGURE 6.13:   Average power consumption of the FPGA device with
different data input/output rates

from 18.4% to 36.3% and 29.0% on average can be achieved by lowering the
clock frequencies of the soft processor when communicating with the low-speed
hardware peripherals.

Figure 6.13 shows the energy dissipation of the FPGA device with different
data input and output rates. Our energy management scheme leads to energy
reduction ranging from 73.3% to 89.9% and 86.8% on average for the different execution scenarios considered in our experiments. Lower data input and
output rates, which imply a longer system idle time between task execution,
lead to more energy savings when our COMA energy management technique
is applied.

6.7   An Implementation Based on TinyOS

In this section, we show the development of *h-TinyOS*, an energy efficient
implementation of the popular component based operating system, *TinyOS*
on soft processors. *h-TinyOS* is written in *nesC* [32], an extension of the ANSI
C language. The architecture of *h-TinyOS* is similar to that of the popular
TinyOS [92] operating system. Thus, *h-TinyOS* bears all the features offered
by component based operating systems discussed in Section 6.2.2.2. Through
the customization of the target soft processor, *h-TinyOS* has a much higher
energy efficiency than the original "pure" software based *TinyOS* operating
system. To our best knowledge, this is the first attempt to port component
based operating systems to soft processors and improve their energy efficiency
by customizing the soft processors.

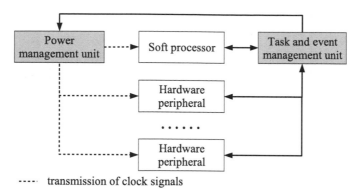

---- transmission of clock signals

FIGURE 6.14:   Hardware architecture of *h-TinyOS*

6.7.1   Hardware Architecture

One major advantage of implementing a component based operating system on a soft processor is that the interactions between the soft processor and the customized hardware peripherals can be clearly described and efficiently controlled through the wiring specifications provided by the component based architecture of the operating system. The "h" in the name of *h-TinyOS* denotes that we tightly couple several customized energy management hardware peripherals to improve the energy efficiency of the operating system. The overall hardware architecture of *h-TinyOS* is shown in Figure 6.14. The functionalities of the customized hardware peripherals for energy management are further discussed in detail in the following paragraphs.

6.7.1.1   Hardware Based Task and Event Management (TEM) Unit

A customized hardware peripheral is designed to perform task and event management in *h-TinyOS*. The management hardware peripheral maintains a task list. When a component posts a task, the information of the task is put into the task list. To better utilize the hardware resources, each task on the task list is associated with a priority number (i.e., the *pri* variable of the FFT computation task shown in Figure 6.16). The management hardware peripheral always selects the task in the task list with the highest priority for execution. The management hardware peripheral also accepts incoming events and invokes the corresponding hardware peripherals to process them. One advantage of the hardware based management approach is efficient task scheduling. The management hardware peripheral can identify the next task for execution within a few clock cycles while the corresponding software implementation usually takes tens of clock cycles. Another advantage is that the soft processor can be turned off when the management hardware peripheral is in operation, thus reducing the energy dissipation of the system. The soft processor is only woken up to perform useful computations. This is in contrast to the state-of-the-art operating systems, including *TinyOS* discussed in Section 6.2.1, where the soft processor has to be active and perform the management functionalities. Therefore, the hardware based management approach reduces the energy dissipation caused by the task and event management.

6.7.1.2 "Explicit" Power Management Unit

The power management unit is a customized hardware peripheral attached to the soft processor for managing the activation states of the various hardware components on the FPGA device. The power management unit provides a software interface, through which *h-TinyOS* can send out commands and "explicitly" control the activation status of the FPGA device. The power management unit is realized through the clock management unit as discussed in Section 6.6.2. Similar to the selective component wake-up unit discussed in Section 6.6.4, each combination of the activation states of the on-chip hardware components of interest is represented by a unique state number. By employing the technique discussed in [87], we allow the application designer to associate a *state* variable with a task. The *state* variable specifies the combination of the activation states of the hardware components (including the soft processor) when the task is under execution. The *state* variable is pushed into the list maintained by the TEM unit discussed above when a task posted by a specific *nesC* component. Before a selected task begins execution, the TEM unit sends out the value of the *state* variable to the power management unit, which will further turn the FPGA device into the desired activation state specified by the *state* variable. Only after that does the selected task actually start execution.

6.7.1.3 Split-phase Operations for Computations Using Hardware Peripherals

We utilize the concept of *split-phase operations* supported by component based operating systems to reduce the power consumption of the system when performing computations in the customized hardware peripherals with long latencies. A component can post a task, which will issue a *command* requesting a *split-phase operation* in the customized hardware peripherals when selected for execution by the management hardware peripheral. The management hardware peripheral will turn off the soft processor and turn on only the hardware peripherals required for computation when the *split-phase* operation is under execution. For a component that needs to be notified of the completion of the *split-phase* operation, the application designer can wire it to the component that requests the *split-phase* operation and implement the *command* function in the wiring interface connecting the two components.

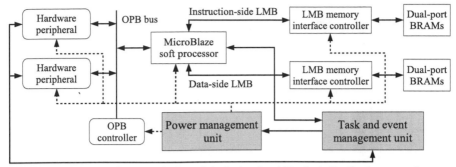

---- denotes the transmission of clock signals within clock distribution network

FIGURE 6.15:   An implementation of *h-TinyOS* on MicroBlaze

6.7.2   Illustrative Application Development

6.7.2.1   Customization of the MicroBlaze Soft Processor

The customization of the MicroBlaze soft processor for running *h-TinyOS* is shown in Figure 6.15. Similar to Figure 6.2, to maintain the real-time property of the operating system, the management hardware components (i.e., the power management unit and the task and event management unit) are attached to the MicroBlaze soft processor through the dedicated FSL bus interfaces. The power management unit controls the clock distribution network on the target FPGA device using the on-chip clock management resources (BUFGCEs and BUFGMUXs as discussed in Figure 6.3).

6.7.2.2   A Case Study

In this section, we develop an FFT embedded application within *h-TinyOS* to demonstrate the effectiveness of our hardware-software co-design technique for improving the energy performance of *h-TinyOS*.

- **Implementation:** We first develop an FFT module for performing FFT (Fast Fourier Transform) computation, the implementation of which is shown in Figure 6.16. The FFT module provides a software wrapper over a customized hardware peripheral, which is realized using an FFT IP (Intellectual Property) core from Xilinx [97] for performing the actual computation. The FFT module provides a *StdControl* interface and an *FFT* interface for interaction with other modules of the operating system. The *StdControl* interface is used to perform initialization during system start-up. The *FFT* interface provides one command *processData()*. Invoking the *processData()* command will put ("post") an FFT computation task into the TEM unit. The application design associates a priority (i.e., *pri*) and an execution status (i.e., *status*) with the FFT computation task. The TEM unit determines when the FFT computation task is ready for execution based on its priority. When the FFT computation task starts execution, it will invoke the wrapped hardware

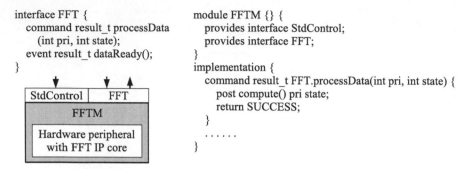

FIGURE 6.16:   Implementation of an FFT module

peripheral in the FFT module to process the data store at the BRAMs. Moreover, before the FFT computation task starts execution, the TEM unit will put the FPGA device into the power status specified by *status* associated with the task.

The FFT interface also provides a *dataReady()* event. For modules that want to be notified about the completion of the FFT computation task, the application designer can "wire" these modules to the FFT module through the *FFT* interface and specify the desired operations within the body of the *dataReady()* event. Once the FFT computation task finishes and signifies the TEM unit the completion of the computation, these operations specified in the *dataReady()* event will be performed.

Since the actual computation of the FFT module is performed using a customized hardware peripheral, *split-phase operations* is employed to improve the energy efficiency of the FFT computation. More specifically, during the execution of the FFT computation task, the MicroBlaze processor is turned off by the clock management unit. Only the FFT hardware peripheral is active and processing the input data. As shown in Table 6.2, this would reduce the energy dissipation of the complete system by 81.0%. When the FFT computation task finishes, if there is any module which needs to notify the completion of the FFT computation task and implements the *dataReady()* event of the *FFT* interface, the TEM unit will automatically wake up the MicroBlaze processor upon receiving the notification from the FFT module. Then, the MicroBlaze soft processor can continue to process the requests from other components.

While *h-TinyOS* can be used for a wide variety of embedded applications, we develop an FFT application based on the FFT module described above in order to demonstrate the effectiveness of *h-TinyOS*. The top-level configuration of the FFT application is illustrated in Figure 6.17. It consists of five major modules: a *Main* module for performing initialization on system startup; a *GPIOM* module for accepting the data from a 32-bit general purpose input/output (GPIO) port (i.e., the data input task); an *FFTM* module for wrapping a hardware peripheral for FFT computation (i.e., the FFT compu-

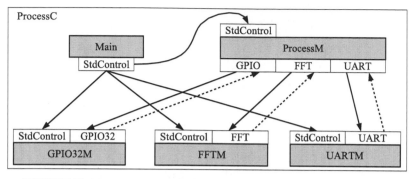

FIGURE 6.17: Top-level configuration *ProcessC* of the FFT computation application

TABLE 6.2: Various activation states for the FFT application shown in Figure 6.17 and their dynamic power consumption

Activation states	Combination of activation states
0	MicroBlaze + GPIO32
1	MicroBlaze + UART (@ 6.25 MHz)
2	All On (for comparison)
3	FFT
4	MicroBlaze + FFT (for comparison)

Activation states	Dynamic power (W)	Reduction
0	1.158	2.5% *
1	0.716	38.7% *
2	1.188	—
3	0.202	81% **
4	1.062	—

∗: compared against state 2; ∗∗ compared against state 4.

tation task); a *UARTM* module for sending out the result data to a personal computer through a serial port (i.e., the data output task); and a *ProcessM* module for wiring the other four modules together to form a complete FFT computation application.

The experimental environment for *h-TinyOS* is similar to that discussed in Section 6.6.6.2. We configure the complete MicroBlaze soft processor on a Xilinx Spartan-3 FPGA prototyping board and measure the power/energy performance of *h-TinyOS* under different execution scenarios (see Section 6.6.6.2 for details regarding the techniques for measuring the power/energy dissipation of the operating system).

• **Experimental results:** The dynamic power consumption of *h-TinyOS* when executing different tasks is shown in Table 6.2. Activation states of 0 and 1 correspond to the states for executing the data input and the data output tasks respectively. With the explicit energy management functionalities provided by *h-TinyOS*, reductions in power consumption of 2.5% and 38.7% are achieved for these two tasks. Specifically, for activation state 2, the power management unit shown in Figure 6.15 can set the operating frequencies of the MicroBlaze processor and the UART hardware peripheral from 6.25 MHz to 50 MHz. The power consumption of the FPGA device when operating with these different operating frequencies is shown in Figure 6.12. By dynamically slowing down the operating frequency of the FPGA device when executing the data output task, power reductions up to 33.2% can be achieved. When the application requires a fixed data output rate determined by the personal computer, the power reduction can directly lead to energy reduction for the data output task.

Activation state 3 denotes the state in which the FFT computation task is executed using *split-phase* operations. In this state, the MicroBlaze processor is turned off while the FFT computation is performed in the customized hardware peripherals. Activation state 2 presents the state that would be set by *TinyOS*, the "pure" software implementation, for executing the FFT computation task. Thus, for the FFT computation task, compared with the "pure" software implementation of *TinyOS*, the hardware-software co-design technique employed by *h-TinyOS* reduces 83.0% of the dynamic power consumption of the FPGA device by explicitly controlling the activation state of the device. Compared with state 4, which represents that activation state that would be set by the MicroC/OS-II implementation discussed in Section 6.6.6.2, *TinyOS* reduces 81.0% of the dynamic power consumption of the FPGA device by allowing the MicroBlaze processor to be turned off during the computation.

Similar to Section 6.6.6.2, we vary the data input and output rates for the data input and data output task. We then actually measure the energy dissipation of the FPGA device using the technique discussed in Section 6.6.6.2. The measurement results are shown in Figure 6.18. *h-TinyOS* achieves reduction on average power consumption ranging from 74.1% to 90.1% and 85.0% on average for these different data rates. Lower data input/output rates, which imply a higher percentage of idle time for the operating system during the operation, lead to more energy reductions.

6.7.3　Analysis of Management Overhead

The software kernel of *h-TinyOS* is less than 100 bytes, as compared against ~400 bytes required by the corresponding "pure" software implementation of *TinyOS*. The reduction of the software kernel size is mainly due to the fact that we let the prioritized task and event queue be maintained by the hardware based task and event management unit tightly attached to the MicroBlaze soft processor. More importantly, with such customized task and event hardware peripheral, *h-TinyOS* can respond to and begin processing the arising tasks

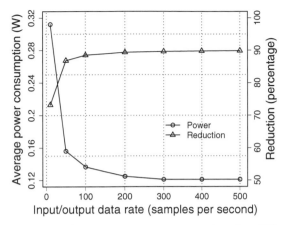

FIGURE 6.18:   Average power consumption for different data input/output rates for *h-TinyOS*

and incoming events between 5 to 10 clock cycles. Also, the time for context switching between different tasks is within 3 clock cycles, which includes one clock cycle for notifying the completion of a task, one clock cycle for identifying the next task with the high priority for execution, and one possible clock cycle for changing the activation state of the FPGA device. This is compared against 20 to 40 clock cycles as required by the "pure" software version of *TinyOS* enhanced with priority scheduling, and the typical hundreds of clock cycles as required by MicroC/OS-II (see Section 6.6.5 for details on the analysis of the management overhead for MicroC/OS-II). Similar to the management overhead analysis discussed in Section 6.6.5, there can be one extra clock cycle for changing the activation states of the FPGA device. However, due to the parallel processing capability offered by customized hardware designs, the task and event management unit is able to find the task in the task queue with the highest priority within one clock cycle, a considerable improvement in time compared with the *for* loops in the corresponding "pure" software implementation. Therefore, considering the above factors, the overall overhead for *h-TinyOS* to manage tasks and events is less than that of "TinyOS". With such a short event responding and task switching time, *h-TinyOS* is very efficient in handling the interactions between the processor and its hardware peripherals as well as coordinating the cooperative processing between them.

6.8   Summary

We proposed a hardware-software co-design based cooperative management technique for energy efficient implementation of real-time operating systems on FPGAs in this chapter. The implementations of two popular real-time operating systems using the proposed hardware-software co-design technique based on a state-of-the-art soft processor, as well as the development of several embedded applications, are shown to demonstrate the effectiveness of our approach.

Chapter 7

Concluding Remarks and Future Directions

7.1 Concluding Remarks

Four major contributions toward energy efficient hardware-software application synthesis using reconfigurable hardware have been presented in this book. For illustrative purposes, we provides implementations of these four techniques based on the state-of-the-art high-level *System Generator* design tool. The development of several practical reconfigurable kernels and applications is also provided in this book. Through these implementations and example designs, we demonstrates that the proposed techniques can lead to significant energy reductions for application development using modern reconfigurable hardware.

7.2 Future Work

As reconfigurable hardware devices are more and more widely used in embedded system development, including many portable devices (e.g., cell phones and personal digital assistants, etc.), energy efficient application development will continue to be a research hot-spot for both academia and industry. The following bullets summarize the potential extensions of the research work presented in this book.

- *Co-simulation:* the high-level co-simulation framework proposed in Chapter 3 gathers the high-level simulation data through the software based cycle-accurate simulation. While the high-level simulation hides a lot of the low-level simulation details and thus is much faster than the traditional low-level hardware simulation, they are still limited by the cycle-by-cycle simulation computation and the computation capability of the personal computers. For example, it is very timing consuming to simulate a MicroBlaze processor running a Linux operating system using the proposed high-level co-simulation framework. One possible extension of the framework to overcome this limitation is hardware based profiling. We can utilize the hardware co-simulation capability provided by *System Generator* to significantly speed up the co-simulation process. The profiling mechanism and hardware circuits are automatically generated and embedded into the design portions to be run on hardware. The profiling data is gathered during the hardware co-simulation process and is transmitted back to the personal computer and is used in the later step for rapid energy estimation.

• *Energy estimation:* for the rapid energy estimation technique presented in Chapter 5, it is desired that the energy estimates can be guaranteed to over-estimate the energy dissipation of the low-level implementations. In addition, providing confidence interval information of the energy estimates would be very useful in practical system development. State-of-the-art synthesizers may perform various transformations on the high-level design models in order to improve the performance of the final hardware designs. This may lead to the deviation of the high-level models and the corresponding low-level implementations. For example, register re-timing is an optimization technique used by many commercial synthesizers (e.g., Synplify [91] and the XST [97]). A register may move to another location of the low-level implementation, rather than the original location specified in the high-level models. Such optimization may introduce a fair amount of inaccuracy in energy estimation. There have been attempts to perform energy estimation based on the synthesis netlists. Many synthesizers can finish the synthesis processes in a reasonable amount of time. We can apply this technique to portions or the complete designs for better estimation accuracy.

• *Energy performance optimization:* for the dynamic programming based energy performance optimization proposed in Chapter 4, various design constraints need to be considered when the end user is targeting a specific configurable hardware device. Examples of such design constraints include execution time and various hardware resource usages (e.g., number of configurable logic blocks and embedded multipliers). When considering these constraints, the energy performance optimization problem would become NP-Complete. No polynomial solutions can be found for this extended version of the problem. However, efficient approximation algorithms and heuristics can still be applied to identify the designs with minimum energy dissipation while satisfying the various constraints.

• *Implementation of operating systems:* we are working to apply the COoperative MAnagement (COMA) technique proposed in Chapter 6 to embedded hard processor cores. We are currently focusing on the IBM PowerPC405 hard processor core integrated in the Xilinx Virtex-4 FX series FPGAs. As is discussed in Chapter 2, the embedded PowerPC has a tight integration with the surrounding configurable logic. The tight integration even allows the end user to customize the decoding process of the PowerPC processor. It is expected that the COMA technique can also significantly reduce the energy dissipation of the real-time operating systems running on the PowerPC hard processor core.

References

[1] Actel, Inc. *http://www.actel.com/*.

[2] M. Adhiwiyogo. Optimal pipelining of i/o ports of the Virtex-II multiplier. In *Xilinx Application Note*, 2003.

[3] Altera, Inc. *http://www.altera.com/*.

[4] Altera, Inc. Cyclone: The lowest-cost FPGA ever. *http://www. altera.com/products/devices/cyclone/cyc-index.jsp*.

[5] Altera, Inc. DSP Builder user guide. *http://www.altera.com/ literature/ug/ug_dsp_builder.pdf*, 2007.

[6] R. Andraka. A survey of CORDIC algorithms for FPGAs. In *Proceedings of the ACM/SIGDA International Symposium on Field Programmable Gate Arrays (FPGA)*, Monterey, California, USA, Feb. 1998.

[7] ARM Limited. Application note: The ARMulator. *http: //infocenter.arm.com/help/topic/com.arm.doc.dai0032f/ AppNote32_ARMulator.pdf*, 2003.

[8] A. Bakshi, V. K. Prasanna, and A. Ledeczi. MILAN: A model based integrated simulation framework for design of embedded systems. In *Proceedings of ACM SIGPLAN/SIGBED Conference on Languages, Compilers, and Tools for Embedded Systems (LCTES)*, 2001.

[9] F. Balarin, M. Chiodo, D. Engeles et al. Hardware-software co-design of embedded systems. The POLIS approach. Kluwer Academic Publisher, 1997.

[10] J. Ballagh, J. Hwang, P. James-Roxby, E. Keller, S. Seng, and B. Taylor. XAPP264: Building OPB slave peripherals using System Generator for DSP. *http://direct.xilinx.com/bvdocs/appnotes/ xapp264.pdf*, July 2004.

[11] J. Becker. Configurable systems-on-chip: Challenges and perspectives for industry and universities. In *Proceedings of Internal Conference on Engineering for Reconfigurable Systems and Algorithms (ERSA)*, 2002.

[12] V. Betz, J. Rose, and A. Marquardt. *Architecture and CAD for Deep-Submicron FPGAs*. Kluwer Academic Publishers, Feb. 1999.

[13] D. Brooks, V. Tiwari, and M. Martonosi. Wattch: A framework for architectural-level power analysis and optimizations. In *Proceedings of International Symposium on Computer Architecture (ISCA)*, 2000.

[14] D. Burger and T. M. Austin. The SimpleScalar tool set, version 2.0. *http://www.simplescalar.com/docs/users_guide_v2.pdf*.

[15] Celoxica, Inc. DK4. *http://www.celoxica.com/products/tools/dk.asp*.

[16] S. Chappell and C. Sullivan. Handel-C for co-processing and co-design of field programmable System-on-Chip. Celoxica, Inc., *http://www.celoxica.com/*, 2004.

[17] S. Choi, J.-W. Jang, S. Mohanty, and V. K. Prasanna. Domain-specific modeling for rapid energy estimation of reconfigurable architectures. *Journal of Supercomputing, Special Issue on Configurable Computing*, 26:259–281, Nov. 2003.

[18] S. Choi, R. Scrofano, and V. K. Prasanna. Energy efficiency of FPGAs and programming processors for various signal processing alogirthms. In *Proceedings of International Conference on Field-Programmable Gate Arrays (FPGA)*, 2003.

[19] J. Cong, Y. Fan, G. Han, A. Jagannathan, G. Reinman, and Z. Zhang. Instruction set extension with shadow registers for configurable processors. In *Proceedings of International Conference on Field-Programmable Gate Arrays (FPGA)*, 2005.

[20] J. Cong, Y. Fan, G. Han, A. Jagannathan, G. Reinman, and Z. Zhang. Instruction set extension with shadow resisters for configurable processors. In *Proceedings of the ACM/SIGDA International Symposium on Field Programmable Gate Arrays (FPGA)*, Feb. 2005.

[21] G. W. Cook and E. J. Delp. An investigation of scalable SIMD I/O techniques with application to parallel JPEG compression. *Journal of Parallel and Distributed Computing (JPDC)*, Nov. 1995.

[22] Crossbow Technology, Inc. *Motes, Smart Dust Sensors*. 2004.

[23] M. Devlin. How to make smart antenna arrays. *Xilinx XCell Journal*, 46, 2003.

[24] C. Dick. The platform FPGA: Enabling the software radio. In *Proceedings of Software Defined Radio Technical Conference and Product Exposition (SDR)*, Nov. 2002.

[25] I. S. I. I. East. Reconfigurable hardware in orbit (RHinO). *http://rhino.east.isi.edu*, 2006.

[26] T. D. Exchange. *http://www.tcl.tk/*.

[27] Express Logic, Inc. ThreadX user guide. *http://www.expresslogic.com*.

[28] M. P. I. Forum. MPI: A message-passing interface standard. *http://www-unix.mcs.anl.gov/mpi/mpi-standard/mpi-report-1.1/mpi-report.htm*, 1995.

[29] P. S. Foundation. Python 2.5 documentation. *http://docs.python.org/*, 2006.

[30] Gaisler Research, Inc. LEON3 user manual. *http://www.gaisler.com/*.

[31] P. Galicki. FPGAs have the multiprocessing I/O infrastructure to meet 3G base station design goals. *Xilinx Xcell Journal*, 45. http://www.xilinx.com/publications/xcellonline/xcell_45/xc_pdf/xc_2dfabric45.pdf, 2003.

[32] D. Gay, P. Levis, R. von Behren, M. Welsh, E. Brewer, and D. Culler. The nesC language: A holistic approach to networked embedded systems. In *Proceedings of International Conference on Programming Language Design and Implementation (PLDI)*, 2003.

[33] A. Gayasen, K. Lee, N. Vijaykrishnan, M. Kandemir, M. Irwin, and T. Tuan. A dual-vdd low power FPGA architecture. In *Proceedings of International Conference on Field Programmable Logic and Its Applications (FPL)*, 2004.

[34] A. Gayasen, Y. Tsai, N. Vijaykrishnan, M. Kandemir, M. J. Irwin, and T. Tuan. Reducing leakage energy in FPGAs using region-constrained placement. In *Proceedings of International Conference on Field-Programmable Gate Arrays (FPGA)*, 2004.

[35] A. Geist, A. Beguelin, J. Dongarra, W. Jiang, R. Manchek, and V. Sunderam. PVM: Parallel virtual machine a users' guide and tutorial for networked parallel computing. MIT Press, 1994.

[36] G. Govindu, L. Zhuo, S. Choi, and V. K. Prasanna. Analysis of high-performance floating-point arithmetic on FPGAs. In *Proceedings of IEEE Reconfigurable Architecture Workshop (RAW)*, 2004.

[37] S. Gupta, M. Luthra, N. Dutt, R. Gupta, and A. Nicolau. Hardware and interface synthesis of FPGA blocks using parallelizing code transformations. In *Proceedings of International Conference on Parallel and Distributed Computing and Systems (ICPADS)*, 2003.

[38] P. Haglund, O. Mencer, W. Luk, and B. Tai. PyHDL: Hardware scripting with python. In *Proceedings of International Conference on Engineering of Reconfigurable Systems and Algorithms (ERSA)*, 2003.

[39] M. Hall, P. Diniz, K. Bondalapati, H. Ziegler, P. Duncan, R.Jain, and J. Granack. DEFACTO: A design environment for adaptive computing technology. In *Proceedings of IEEE Reconfigurable Architectures Workshop (RAW)*, 1999.

[40] M. Hammond. Python for windows extensions. *http://starship.python.net/crew/mhammond*, 2006.

[41] R. Hartenstein and J. Becker. Hardware/software co-design for data-driven Xputer-based accelerators. In *Proceedings of International Conference on VLSI Design: VLSI in Multimedia Applications*, 1997.

[42] S. Haykin. *Adaptive Filter Theory (3rd Edition)*. Prentice Hall, 1991.

[43] R. Hogg and E. Tanis. *Probability and Statistical Inference (6th Edition)*. Prentice Hall, 2001.

[44] P. Holmberg. Domain-specific platform FPGAs. *FPGA and Structure ASIC Journal. http://www.fpgajournal.com/articles/platform_xilinx.htm*, 2003.

[45] M. Horowitz, T. Indermaur, and R. Gonzalez. Low-power digital design. In *Proceedings of IEEE Symposium on Low Power Electronics*, 1994.

[46] J. Hwang, B. Milne, N. Shirazi, and J. Stroomer. System level tools for dsp in fpgas. In *Proceedings of International Conference on Field Programmable Logic and its applications (FPL)*, 2001.

[47] Impulse Accelerated Technology, Inc. CoDeveloper. *http://www.impulsec.com/*.

[48] International Business Machines (IBM). *http://www.ibm.com/*.

[49] International Business Machines (IBM). CoreConnect bus architecture. *http://www-01.ibm.com/chips/techlib/techlib.nsf/productfamilies/CoreConnect_Bus_Architecture*.

[50] P. James-Roxby, P. Schumacher, and C. Ross. A single program multiple data parallel processing platform for FPGAs. In *Proceedings of IEEE International Symposium on Field-Programmable Custom Computing Machines (FCCM)*, 2004.

[51] Y. Jin, K. Ravindran, N. Satish, and K. Keutzer. An FPGA-based soft multiprocessor system for ipv4 packet forwarding. In *Proceedings of IEEE Workshop on Architecture Research using FPGA Platforms (WARFP)*, 2005.

[52] A. K. Jones, R. Hoare, and D. Kusic. An FPGA-based VLIW processor with custom hardware execution. In *Proceedings of ACM International Symposium on Field Programmable Gate Arrays (FPGA)*, 2005.

[53] Keithley Instruments, Inc. *http://www.keithley.com/*.

[54] P. B. Kocik. PacoBlaze - A synthesizable behavioral verilog PicoBlaze clones. *http://bleyer.org/pacoblaze/*, 2007.

[55] D. Lampret, C.-M. Chen, M. Mlinar, and et al. OpenRISC 1000 architecture manual. *http://www.opencores.org/*.

[56] J. J. Larbrosse. *MicroC/OS-II the Real-Time Kernel (2nd Edition)*. CMP Books, 2002.

[57] C. Lee, M. Potkonjak, and W. H. Mangione-Smith. MediaBench: A tool for evaluating and synthesizing multimedia and communications systems. In *Proceedings of International Symposium on Microarchitecture (MICRO)*, 1997.

[58] M. Lee, W. Liu, and V. K. Prasanna. Parallel implementation of a class of adaptive signal processing applications. *Algorithmica*, (30):645–684, 2001.

[59] F. Li, Y. Lin, and L. He. FPGA power reduction using configurable dual-vdd. In *Proceeedings of International Conference on Field-Programmable Gate Arrays (FPGA)*, 2004.

[60] MathWorks, Inc. *http://www.mathworks.com/*.

[61] S. McCloud. Algorithmic C synthesis optimizes ESL design flows. *Xilinx Xcell Journal*, 50, 2004.

[62] Mentor Graphics. Catapult C synthesis. *http://www.mentor.com/products/c-based_design/*.

[63] Mentor Graphics, Inc. *http://www.mentor.com/*.

[64] J. Mitola. The software radio architecture. *IEEE Communications Magazine*, 33(5):26–38, 1995.

[65] V. J. Mooney and D. M. Blough. A hardware-software real-time operating system framework for SoCs. *IEEE Design and Test of Computers*, Nov/Dec 2002.

[66] Nu Horizons Electronics, Inc. *http://www.nuhorizons.com/*.

[67] Open SystemC Initiative. *http://www.systemc.org/*.

[68] J. Ou, S. Choi, and V. K. Prasanna. Performance modeling of reconfigurable soc architectures and energy-efficient mapping of a class of applications. In *Proceedings of IEEE International Symposium on Field Customizable Computing Machines (FCCM)*, 2003.

[69] J. Ou and V. K. Prasanna. Parameterized and energy efficient adaptive beamforming using system generator. In *Proceedings of International Conference on Acoustics, Speech, and Signal Processing (ICASSP)*, 2004.

[70] J. Ou and V. K. Prasanna. PyGen: A MATLAB/Simulink based tool for synthesizing parameterized and energy efficient designs using FPGAs. In *Proceedings of IEEE International Symposium on Field-Programmable Custom Computing Machines (FCCM)*, 2004.

[71] J. Ou and V. K. Prasanna. Rapid energy estimation of computations on FPGA based soft processors. In *Proceedings of IEEE International System-on-a-Chip Conference (SoCC)*, 2004.

[72] J. Ou and S. P. Seng. A high-level modeling environment based approach for the study of configurable processor systems. In *Proceedings of IEEE International Conference on Microelectronic Systems Education (MSE)*, 2007.

[73] K. V. Palem, S. Talla, and W.-F. Wong. Compiler optimizations for adaptive EPIC processors. In *Proceedings of Workshop on Embedded Software*, 2001.

[74] T. P. Plaks and G. M. Megson. Engineering of reconfigurable hardware/software objects. *Journal of Supercomputing*, 19(1):5–6, 2001.

[75] K. K. W. Poon, S. J. E. Wilton, and A. Yan. A detailed power model for field-programmable gate arrays. *ACM Transactions on Design Automation of Electronic Systems (TODAES)*, 10(2):279–302, 2005.

[76] V. K. Prasanna and Y. Tsai. On synthesizing optimal family of linear systolic arrays for matrix multiplication. *IEEE Transactions on Computers*, 40(6), 1991.

[77] W. Press, B. Flannery, S. Teukolsky, and W. Vetterling. *Numerical Recipes in C: The Art of Scientific Computing (Second Edition)*. Cambridge University Press, 2002.

[78] A. Raghunathan, N. K. Jha, and S. Dey. *High-Level Power Analysis and Optimization*. Kluwer Academic Publishers, 1998.

[79] D. Rakhmatov and S. Vrudhula. Hardware-software bipartitioning for dynamically reconfigurable systems. In *Proceedings of International Conference on Hardware Software Codesign (CODES)*, 2002.

[80] J. Razavilar, F. Rashid-Farrokhi, and K. J. R. Liu. Software radio architecture with smart antennas: A tutorial on algorithms and complexity. *IEEE Journal on Selected Area in Communication (JSAC)*, 17(4):662–676, 1999.

[81] R. Scrofano, L. Zhuo, and V. K. Prasanna. Area-efficient evaluation of arithmetic expressions using deeply pipelined floating-point cores. In *Proceedings of the 2005 International Conference on Engineering of Reconfigurable Systems and Algorithms*, June 2005.

[82] M. Shalan and V. J. Mooney. Hardware support for real-time embedded multiprocessor system-on-a-chip memory management. In *Proceedings of International Symposium on Hardware/Software Codesign (CODES)*, 2002.

[83] L. Shannon and P. Chow. Simplifying the integration of processing elements in computing systems using a programmable controller. In *Proceedings of IEEE International Symposium on Field-Programmable Custom Computing Machines (FCCM)*, 2005.

[84] C. Shi, J. Hwang, S. McMillan, A. Root, and V. Singh. A system level resource estimation tool for FPGAs. In *Proceedings of International Conference on Field Programmable Logic and Its Applications (FPL)*, 2004.

[85] A. Sinha and A. Chandrakasan. JouleTrack: A web based tool for software energy profiling. In *Proceedings of Design Automation Conference (DAC)*, 2001.

[86] C. Souza. IP columns support application specific FPGAs. *EE Times.* *http://www.eetimes.com/story/OEG20031208S0061*, 2003.

[87] V. Subramonian, H.-M. Huang, and S. Datar. Priority scheduling in TinyOS : A case study. In *Technical Report*. Washington University, 2003.

[88] F. Sun, S. Ravi, A. Raghunathan, and N. K. Jha. A scalable application-specific processor synthesis methodology. In *Proceedings of International Conference on Computer-Aided Design (ICCAD)*, 2003.

[89] Synfora, Inc. *http://www.synfora.com/*.

[90] Synopsys, Inc. SMART simulation model. *http://www.synopsys.com/products/lm/doc/smartmodel.html*, 2005.

[91] Synplicity, Inc. *http://www.synplicity.com*.

[92] TinyOS. TinyOS documentation. *http://www.tinyos.net/tinyos-1.x/doc/*.

[93] T. Tuan and B. Lai. Leakage power analysis of a 90nm FPGA. In *Proceedings of IEEE Custom Integrated Circuits Conference (CICC)*, 2003.

[94] J. Villarreal, D. Suresh, G. Stitt, F. Vahid, and W. Najjar. Improving software performance with configurable logic. *Kluwer Journal on Design Automation of Embedded Systems*, 7:325–339, 2002.

[95] J. wook Jang, S. Choi, and V. K. Prasanna. Energy-efficient matrix multiplication on FPGAs. *IEEE Transactions on Very Large Scale Integrated Circuits and Electronic Systems (TVLSI)*, 13(11):1305–1319, Nov. 2005.

[96] Y. Xie and W. Wolf. Allocation and scheduling of conditional task graph in hardware/software co-synthesis. In *Proceedings of Design, Automation, and Test in Europe*, 2001.

[97] Xilinx, Inc. *http://www.xilinx.com/*.

[98] Xilinx, Inc. EasyPath series. *http://www.xilinx.com/products/ silicon_solutions/fpgas/easypath/index.htm*.

[99] Xilinx, Inc. Xilinx application note: Virtex-II series and xilinx ISE 4.1i design environment. *http://www.xilinx.com*, 2001.

[100] Xilinx, Inc. PicoBlaze 8-bit embedded microcontroller user guide for Spartan-3, Virtex-II, and Virtex-II Pro FPGAs. *http://www.xilinx. com/bvdocs/userguides/ug129.pdf*, 2003.

[101] Xilinx, Inc. Two flows for partial reconfiguration: Module based or difference based (XAPP290). In *Xilinx Application Notes. http:// www.xilinx.com/bvdocs/appnotes/xapp290.pdf*, Sept. 2004.

[102] Xilinx, Inc. Xilinx announces acquisition of triscend corp. *http://www.xilinx.com/prs_rls/xil_corp/0435_triscend_ acquisition.htm*, 2004.

[103] Xilinx, Inc. Web power analysis tools. *http://www.xilinx.com/ power/*, 2006.

[104] Xilinx, Inc. AccelDSP synthesis tool. *http://www.xilinx.com/ise/ dsp_design_prod/acceldsp/index.htm*, 2007.

[105] Xilinx, Inc. PlanAhead methodology guide (release 9.2). *http://www. xilinx.com/ise/planahead/PlanAhead_MethodologyGuide.pdf*, 2007.

[106] Xilinx, Inc. Power analysis of 65nm Virtex-5 FPGAs. *http://www. xilinx.com/bvdocs/whitepapers/wp246.pdf*, 2007.

[107] Xilinx, Inc. Spartan-3 Generation FPGA User Guide. *http:// direct.xilinx.com/bvdocs/userguides/ug331.pdf*, 2007.

[108] Xilinx, Inc. System Generator for DSP user guide. *http://www. xilinx.com/support/sw_manuals/sysgen_bklist.pdf*, 2007.

[109] Xilinx, Inc. Virtex-5 user guide. *http://direct.xilinx.com/ bvdocs/userguides/ug190.pdf*, 2007.

[110] Xilinx, Inc. Virtex-II Pro and Virtex-II Pro X FPGA user guide. *http: //direct.xilinx.com/bvdocs/userguides/ug012.pdf*, 2007.

[111] Xilinx, Inc. Xilinx demonstrates industrys first scalable 3-d graphics hardware accelerator for automotive applications. *http://www. xilinx.com/prs_rls/2007/end_markets/0703_xylon3dCES.htm*, 2007.

[112] Xilinx, Inc. Xilinx power estimator user guide. *http://www.xilinx. com/products/design_resources/power_central/ug440.pdf*, 2007.

[113] Xilinx, Inc. Xilinx SDR radio kit wins 2006 portable design editor's choice award. *http://www.xilinx.com/prs_rls/2007/xil_corp/ 0733_pdawards.htm*, 2007.

[114] Xilinx, Inc. Xilinx Spartan-3E FPGAs enable JVC's GY-HD250. *http://digitalcontentproducer.com/hdhdv/prods/ xilinx_hdv_11212006/*, 2007.

[115] Xilinx Inc. Xilinx virtex-ii pro web power tool version 8.1.01. *http: //www.xilinx.com/cgi-bin/power_tool/power_Virtex2p*, 2007.

[116] W. Ye, N. V. Krishnan, M. Kandemir, and M. J. Irwin. The design and use of SimplePower: A cycle-accurate energy estimation tool. In *Proceedings of Design Automation Conference (DAC)*, 2000.

[117] Z. A. Ye, A. Moshovos, S. Hauck, and P. Banerjee. CHIMAERA: A high-performance architecture with a tightly-coupled reconfigurable functional unit. In *Proceedings of International Symp. on Computer Architecture (PISCA)*, 2000.

[118] M. Hubner, K. Paulsson, and J. Becker. Parallel and flexible multiprocessor system-on-chip for adaptive automotive applications based on Xilinx MicroBlaze soft-cores, In *Reconfigurable Architectures Workshop (RAW)*, 2005.

[119] L. Zhuo and V. K. Prasanna. Scalable and modular algorithms for floating-point matrix multiplication on FPGA. In *Proceedings of IEEE International Parallel and Distributed Processing Symposium (IPDPS)*, 2004.

Index

Printed and bound by CPI Group (UK) Ltd, Croydon, CR0 4YY

18/10/2024

01776243-0002